Filter Banks and Audio Coding

Gerald Schuller

Filter Banks and Audio Coding

Compressing Audio Signals Using Python

 Springer

Gerald Schuller
Ilmenau University of Technology
Ilmenau, Germany

ISBN 978-3-030-51251-4   ISBN 978-3-030-51249-1   (eBook)
https://doi.org/10.1007/978-3-030-51249-1

This Springer imprint is published by the registered company Springer Nature Switzerland AG
The registered company address is: Gewerbestrasse 11, 6330 Cham, Switzerland

Preface

This book has been developed from our lecture "Audio Coding", together with Prof. Karlheinz Brandenburg, at Ilmenau University of Technology, Germany.

Audio coding was started several decades ago with the goal to transmit music over the then much slower computer networks, the digital ISDN telephone network at $64\,\text{kb/s}$, and for the then-evolving digital radio and TV broadcasting standards, in the 1980s and 1990s.

Some of the driving forces then were the group of K. Brandenburg at Fraunhofer IIS in Erlangen, Germany [1] and the group of J. Johnston at then AT&T Bell Labs, in Murray Hill, New Jersey, USA [2]. They collaborated, and the German side participated in the EU project called EUREKA 147 for the development of digital radio or audio broadcasting system [3].

A then-developing popular application was the storage of music on the much smaller computer hard drives, for which compression was essential to make it practical. This led to portable music devices, with even more constrained memory, the MP3 players. In the 2000s, smartphones evolved, combining wireless phones and MP3 players, most notably the iPhone in 2007, and the iTunes music library, which made it possible to buy and download music wirelessly, without the need to connect to a computer with a fixed Internet connection. Around the same time, the Internet and wireless connections were becoming fast enough to make Internet radio and streaming more popular. The speed of the wireless connections was previously a bottleneck for music streaming, which again shows the necessity for good audio compression. The speed of wireless and wired Internet connections nowadays makes streaming the most popular mode for listening to music, still relying on good audio compression. This is particularly true because the more popular music streaming becomes, the more network bandwidth is used, and hence it becomes more important that each listener uses as little bandwidth as possible for good audio quality.

With high speed and low delay networks and increased computing power also came applications like high audio quality teleconferencing. There, the encoding/decoding delay becomes important. Traditional audio encoding/decoding has too much delay for it; hence, low delay audio coders were developed for that purpose in the 2000s.

This book focuses on the fundamentals or basics of audio coding and uses examples in the programming language Python, also as an algorithmic description. Python is an open-source very high-level language, which makes it suitable as an algorithmic description language, and as a basis for your own experiments. This follows the motto "only what one can create, one fully understands". Python is also a full programming language, allowing working audio coders to be built with it. They can also be a basis or an example for implementations for engineers. This prepares the reader for more advanced techniques in audio coding (for instance, parametric bandwidth extension or parametric spatial audio coding).

The fundamentals of audio coding are at the highest level the two principles of "redundancy reduction" and "irrelevance reduction". Redundancy means the statistical dependencies of a signal, and it is taken advantage of by using models of the signal. Irrelevance means the information that the receiver, here the human ear, cannot detect. It is taken advantage of by using models of hearing, psycho-acoustic models.

The book is structured such that it first covers the signal models, in the form of time/frequency decompositions and filter banks, like the Discrete Cosine Transform, the Modified Discrete Cosine Transform, or the Pseudo Quadrature Mirror Filter bank, and predictors.

Then it covers models of hearing, with a psycho-acoustic model and an example implementation.

Finally, it shows several complete coders, with Python implementations. First a perceptual subband coder, then a predictive lossless coder, a scalable lossless audio coder, and a low delay audio coder using a psycho-acoustic pre- and post-filter.

The code examples are made such that they work with both, Python2 and Python3. The software accompanying this book can be downloaded from `https://github.com/TUIlmenauAMS/Python-Audio-Coder`, for instance, in a terminal shell using `git clone https://github.com/TUIlmenauAMS/Python-Audio-Coder`.

Ilmenau, Germany   Gerald D. T. Schuller
May 2020

References

1. K. Brandenburg, Evaluation of quality for audio encoding at low bit rates, in *82nd AES Convention* (1987)

2. J. Johnston, Transform coding of audio signals using perceptual noise criteria. IEEE J. Sel. Areas Commun. **6**(2), 314–323 (1988)

3. "EUREKA-147 – Digital Audio Broadcasting", https://pdfs.semanticscholar.org/96ba/d214344f3ae6b43ac7293115bcf274f3f7af.pdf

Contents

Contents

About the Author

Gerald D. T. Schuller has been a full professor at the Institute for Media Technology of the Technical University of Ilmenau since 2008. His research interests are in filter banks, audio coding, music signal processing, and deep learning for multimedia. He was a Member of Technical Staff at Bell Laboratories, Murray Hill, New Jersey, from 1998 to 2001. There, he worked in the Multimedia Communications Research Laboratory, among others, on audio coding for satellite radio.

He then returned to Germany to become head of the Audio Coding for Special Applications group of the Fraunhofer Institute for Digital Media Technology in Ilmenau, Germany, from January 2002 until 2008, where he worked, among others, on MPEG AAC-ELD, Ultra Low Delay audio coding, and scalable lossless audio coding.

1 Filter Banks

1.1 Introduction

The goal of audio coding is to obtain a high compression ratio with a good audio quality after decompression or decoding. Hence both irrelevance reduction (the removal of information the receiver, here the ear, cannot detect) and redundancy reduction (the statistical dependencies in the signal) [1–3] should be applied as much as possible. The most straightforward way to apply both principles is to use subband coding, which uses a decomposition of the audio signal into subbands. The audio signal consists of samples with a given sampling rate. For instance, if it comes from an audio CD, it has a sampling rate of 44,100 samples/sec or 44.1 kHz.

The purpose of filter banks is to perform this split or decomposition of the signal into subbands. For instance, if the signal is an audio signal, sampled at 44.1 kHz, which is to split into 1024 subbands with equal bandwidth, then subband 0 contains the signal frequency components from 0 to 21.53 Hz, band 1 contains 21.53 to 43.07, and band 1023 contains 22,028.47 to 22,050 Hz.

The filter bank which performs this decomposition is located in the encoder, and is called the "analysis filter bank". The decoder needs the reverse function. It needs a filter bank to reconstruct the original audio signal out of the subband signals. This filter bank in the decoder is called the "synthesis filter bank".

Our requirements for the filter bank are:

(1) a good frequency separation. We would like one frequency component only to appear in one subband, or if it appears also in other subbands then at a level as low as possible. Otherwise we would encode the same component several times in different subbands, wasting bits.

(2) Critical sampling. Assume the filter bank has N subbands. We don't want to increase the total number of samples by N. For that reason we would like to downsample each subband signal by a factor of N, meaning we only keep every N-th sample, so that the total number of samples stays constant.

(3) The filter bank should have perfect reconstruction property, despite the critical sampling. This means if we connect the analysis filter bank from the encoder directly with the synthesis filter bank from the decoder, without any quantization or processing, the original audio signal should be reconstructed by the synthesis filter bank. This also means the filter bank does not produce any artifact on its own.

© Springer Nature Switzerland AG 2020
G. Schuller, *Filter Banks and Audio Coding*,
https://doi.org/10.1007/978-3-030-51249-1_1

(4) The filter bank should have an efficient implementation. This is important because in audio coding there is usually a high number of subbands used, which would lead to a complex hardware implementation otherwise.

(5) The filter bank should be able to be switched to different numbers of subbands during processing, while maintaining perfect reconstruction and also good separation of subbands. This is important to adapt the filter bank to different signal statistics with different time/frequency trade-offs. This is important for very non-stationary signals, like attacks from castanets, to obtain a high enough time resolution to use temporal masking effects of the ear (the backward masking effects), to avoid the so-called pre-echoes. It should be able to take advantage of the more stationary nature of many audio signals by switching to a higher number of narrower subbands with long filter impulse responses.

Figure 1.1 shows the basic structure of the analysis (on the left) and synthesis filter bank (on the right) (see also [4, 5]). The boxes symbolize the filters with impulse responses $h_k(n)$ in the analysis or $g_k(n)$ in the synthesis filter bank, for $k = 0, \ldots, N-1$. The boxes with $\downarrow N$ mean downsampling by N, $\uparrow N$ means upsampling by N to obtain the original sampling rate by inserting $N-1$ zeros after each sample. $y_k(m)$ are the subband signals. Since they are downsampled and hence have a different "time basis", the time index is m instead of n. The index m can be seen as block index, if the audio signal is divided into blocks of length N. The signal flow is along the lines ("wires") from left to right.

Analysis   Synthesis

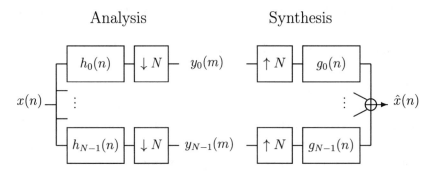

Figure 1.1: An N-band-filter bank with critical sampling. Perfect reconstruction means that $\hat{x}(n) = x(n - n_d)$, where n_d is the system delay. $\downarrow N$ means downsampling by N, meaning only every N-th sample is kept, hence reducing the sampling rate accordingly. $\uparrow N$ means upsampling by N to obtain the original sampling rate. This is done by inserting $N-1$ zeros after each sample.

The output of the analysis filter bank is also called a "time-frequency representation", because it is a decomposition into different frequency bands ("frequency"), and it still features a dependency on the time, even though at a coarser resolution due to the length of the filter impulse responses, which can be seen as performing an averaging over their length, and the following downsampling.

How can the requirements for the filter bank be fulfilled?

(1) a good frequency separation. Ideally we would have subband filters which would let frequencies in the corresponding subband pass unaltered (the passband), and frequencies outside the passband be attenuated to zero. But this requirement leads to filters with infinitely long impulse responses and an infinite delay, and hence are not realistic.

(2) critical sampling is fulfilled since downsampling factor is equal to the number of bands N.

(3) perfect reconstruction. How can perfect reconstruction of the original audio signals out of the downsampled subbands be achieved, despite the filtering and downsampling?

To answer those questions, we start with FIR filter structures and then analysing sampling and downsampling.

1.2 Frequency Domain Transforms and Notation

We will use lowercase letters for time-domain signals, like $x(n)$, and uppercase letters for transform domain signals, like $X(\Omega)$ or $X(z)$. We will use boldface letters to denote vectors or matrices in both time domain and transform domain, like $\mathbf{x}(m)$ or $\mathbf{X}(z)$. The conjugate complex operation is symbolized with a superscript asterisk, like $x^*(n)$. For transform domain signals, like $X(z)$ or $X(e^{j\omega})$ the asterisk ($X^*(z)$ or $X^*(e^{j\omega})$) denotes the conjugate complex operation on their coefficients only, not on their arguments (z or $e^{j\omega}$). The overline ($\overline{X(z)}$ or $\overline{X^*(e^{j\omega})}$) denotes the conjugate complex operation on the final result, meaning including the arguments. $E(x)$ is the "expectation" of $x(n)$, for our purposes the average of $x(n)$. The sign ":=" means "defined as", for instance, $a := b$. If we have different sampling rates, we will use n as time index at the higher sampling rate, and m as time index at the lower sampling rate. Downsampling a signal $x(n)$ at phase i (with $0 <= i <= N - 1$) is symbolized with a superscript $\downarrow N$ and subscript i as

$$x_i^{\downarrow N}(m) := x(mN + i).$$

Its z-transform is written as

$$X_i^{\downarrow N}(z) = \sum_{m=0}^{\infty} x(mN + i) \cdot z^{-m}$$

Upsampling a signal $y(m)$ with phase i is written as

$$y_i^{\uparrow N}(n) := \begin{cases} y(m) & \text{if} \quad n = mN + i \\ 0 & \text{else} \end{cases}$$

To describe sampling mathematically we will also use a function which we call delta train. It is an infinite sequence with zeros, where every N'th sample is 1,

$$\delta_N(n) = \begin{cases} 1 & \text{if} \quad n = mN \\ 0 & \text{else} \end{cases} \tag{1.1}$$

Observe that this implies phase 0 (starting sampling at position 0). For other phases n_0 we simply shift the delta train,

$$\delta_N(n - n_0)$$

Its Discrete Time Fourier Transform (see below) is a Dirac function train at frequencies $2\pi/N$,

$$\Delta_{\frac{2\pi}{N}}(\omega) = \sum_{k=0}^{N-1} \delta(\omega - k\frac{2\pi}{N}) \tag{1.2}$$

To design and analyse filter banks, we need to use different types of transforms. They are listed in the following (see also [6]). First we have different types of Fourier Transforms. We will use lowercase letters to indicate a time-domain signal and corresponding uppercase letters to indicate its frequency or transform domain signal.

- The **Fourier Series** is the oldest frequency domain transform. It assumes a **periodic time-continuous** signal $x(t)$ with period T to compute coefficients $X(k)$ for the k'th harmonic frequency of the signal. It is defined as

$$X(k) = \int_{-T/2}^{T/2} x(t)e^{-j\frac{2\pi}{T} \cdot kt}dt$$

with k integer as the frequency index. Observe that the Fourier series coefficients $c(k)$ are **discrete in frequency**, because the signal is periodic, and **infinite** in k. Example: A signal with a period of $1/100$ s has a fundamental frequency of ±100 Hz, with coefficient $X(0)$, and harmonic frequencies at multiple of ±100 Hz, with coefficients $X(k)$ with k integer. The inverse is

$$x(t) = \sum_{k=-\infty}^{\infty} X(k) \cdot e^{j\frac{2\pi}{T} \cdot kt}$$

- The **Fourier Transform** is again for time-continuous signals $x(t)$, but now the signal is not necessarily periodic. Instead it can be any signal of **infinite extent**. It is defined as [7]

$$X(f) = \int_{-\infty}^{\infty} x(t)e^{-j2\pi ft}dt$$

or

$$X(\omega) = \int_{-\infty}^{\infty} x(t)e^{-j\omega t}dt$$

Here we assume that the signal is such that the integral converges (is finite). Observe that the Fourier Transform is now **continuous in frequency** f or ω (because our signal is no longer periodic) and is **infinite** in extent. In the latter equation the frequency variable is $\omega = 2\pi f$, a continuous angular frequency. The inverse is

$$x(t) = \int_{-\infty}^{\infty} X(f)e^{-j2\pi ft}df$$

or

$$x(t) = \frac{1}{2\pi} \int_{-\infty}^{\infty} X(\omega)e^{-j\omega t}d\omega$$

- The **Discrete Time Fourier Transform (DTFT)** works on **discrete time** (sampled) signals $x(n)$, where n is integer and the time sample index, and where the signal is again of **infinite extent** [6, 8]. If we sample the analog signal $x(t)$ with a sampling interval T, and formulate the sampling as multiplication with dirac pulses at the sampling instances (more below), with sampling frequency $f_s = 1/T$, then the integral of Fourier Transform becomes a sum. The time variable t becomes nT, and in its exponential function of the transform kernel we get the expression $2\pi f nT = 2\pi n f/f_s$. Hence we get a new frequency variable $\Omega = 2\pi f/f_s$, which is the **normalized** angular frequency. There the sampling frequency f_s corresponds to the normalized frequency 2π,

$$X(e^{j\Omega}) = \sum_{n=\infty}^{\infty} x(n)e^{-j\Omega \cdot n}$$

Since our time signal is infinite, we get a **continuous frequency** variable Ω, and since the time signal is discrete, Ω is also 2π **periodic**.

The Inverse Discrete Time Fourier Transform is

$$x(n) = \frac{1}{2\pi} \cdot \int_{\Omega=-\pi}^{\pi} X(e^{j\Omega})e^{j\Omega n}d\Omega$$

The **Parseval's theorem** [6, 8] for the DTFT states that the signal power we compute in the time domain is identical to the signal power we compute in the frequency domain,

$$E(x^2) = \frac{1}{2\pi} \cdot \int_{\Omega=-\pi}^{\pi} |X(e^{j\Omega})|^2 d\Omega \tag{1.3}$$

- The **Discrete Fourier Transform (DFT)** works again on **discrete time** signals $x(n)$ but now for finite length signals, which are assumed to be **periodic in time** beyond this length (periodic with period of N samples) [6]. It is defined as

$$X(k) = \sum_{n=0}^{N-1} x(n)e^{-j2\pi/N \cdot k \cdot n}$$

It is **discrete in frequency**, because of the periodicity in time. Hence the frequency index is now k, an integer between 0 and $N-1$. It is **periodic in frequency**, with period N, because the signal is discrete in time. $k = N$ corresponds to the normalized frequency 2π. The inverse DFT is

$$x(n) = \frac{1}{N} \sum_{k=0}^{N-1} X(k)e^{j2\pi/N \cdot k \cdot n}$$

- The **z-Transform** can be seen as a generalization of the DTFT, because we can obtain it by replacing the exponential function $e^{j2\pi/N \cdot k \cdot n}$, which is a complex number on the unit circle in the complex plane, by any complex number z. Hence it is also for **discrete time** signals of **infinite extent**. Since we are dealing with causal signals (which have a beginning in time, for instance at $n = 0$), we use the so-called one-sided z-Transform, defined as [6, 8]

$$X(z) = \sum_{n=0}^{\infty} x(n) z^{-n}$$

We can see that we obtain the DTFT if we have a causal signal and if we replace z by $e^{j2\pi/N \cdot k \cdot n}$. Again we have a **continuous periodic frequency** variable, now periodic on the unit circle in the z-domain. This transform has the advantage that we can also determine, for instance, if a system or a signal is stable or if it has damping on it (in that case the signal or system would have poles in the z-domain inside the unit circle).

The inverse z-transform is

$$x(n) = \frac{1}{2\pi j} \oint_c X(z) z^{n-1} dz$$

where C is a closed contour in the complex z-plane which contains all poles of $X(z)$ in its inside, and its path is followed counter-clockwise in the mathematically positive sense. If we have a causal, stable z-transform $X(z)$, then all poles are inside the unit circle, hence our contour C can be the unit circle around the origin, and we can set $z = e^{j\omega}$ for $0 \le \omega \le 2\pi$. We can now apply an integration variable substitution

$$\frac{dz}{d\omega} = je^{j\omega}$$

$$\implies dz = d\omega \cdot je^{j\omega}$$

(using the Leibnitz notation). Now the contour integral becomes

$$x(n) = \frac{1}{2\pi j} \int_{\omega=0}^{\pi} X(e^{j\omega}) e^{j\omega(n-1)} d\omega \cdot je^{j\omega}$$

$$= \frac{1}{2\pi} \int_{\omega=0}^{\pi} X(e^{j\omega}) e^{j\omega n} d\omega$$

Now we can see that the last line is identical to the inverse DTFT, which means that for causal stable signals or systems the inverse z-transform becomes identical to the inverse DTFT.

Observe that we have transform pairings of periodic time—discrete frequency, and discrete time—periodic frequency.

1.3 Linear Filters

We assume the filters $h_k(n)$ and $g_k(n)$ we saw to be linear and time-invarying (LTI) systems. For an LTI system S we have the following basic rules or conditions [6]. For linearity

$$S\left(x(n)+y(n)\right) = S(x(n)) + S(y(n))$$
$$S(c \cdot x(n)) = c \cdot S(x(n))$$

which means we can draw sums and constant multiplications out of our system. For time-invariance we have

$$S(x(n+m)) = S\left(x(n)\right)(n+m)$$

which means the behaviour of our system is independent of the time at which we feed the input (a shifted input leads to a correspondingly shifted output). We can now use an elementary "atom" for our signals, the delta function, which is a 1 at time $n=0$ and 0 at all other times,

$$\delta(n) = \begin{cases} 1 & \text{if } n = 0 \\ 0 & \text{otherwise} \end{cases}$$

We can now write any signal $x(n)$ as a weighted sum of shifted delta functions $\delta(n-m)$,

$$x(n) = \sum_{m=-\infty}^{\infty} x(m) \cdot \delta(n-m)$$

observe that the $x(m)$ in the sum can be seen as fixed weights of the shifted delta functions. We now apply our LTI system S to this signal, and use the LTI rules to simplify it,

$$S(x(n)) = S\left(\sum_{m=-\infty}^{\infty} x(m) \cdot \delta(n-m) \right)$$

$$= \sum_{m=-\infty}^{\infty} x(m) \cdot S(\delta(n-m))$$

$$= \sum_{m=-\infty}^{\infty} x(m) \cdot S(\delta(n))(n-m)$$

Here we can see that the resulting output of our LTI system S to the input $x(n)$ is computed using the response of the system to the delta function $\delta(n)$. Since the delta function is basically an "impulse", we call the response to the delta function the "impulse response". Here we can write it as

$$h(n) := S(\delta(n))$$

and with definition we can rewrite the output of our LTI system S as

$$S(x(n)) = \sum_{m=-\infty}^{\infty} x(m) \cdot h(n-m) \tag{1.4}$$

Here we can see that the LTI system is completely specified by its impulse response! Equation (1.4) is also called a "convolution". The short notation is an asterisk,

$$S(x(n)) = x(n) * h(n)$$

This is also how filters are implemented in the time domain, as a convolution of the signal with the filters impulse response. If the impulse response is infinite in length, we call the filter "Infinite Impulse Response" (IIR) filter, if it is finite, we call it a "Finite Impulse Response" (FIR) filter.

Our goal for the filter is that they attenuate certain frequencies in our signal, and let other frequencies pass. To see the effect of our filtering we can apply the Discrete Time Fourier Transform (DTFT), see next Sect. 1.4, to our filtered signal

$$y(n) = x(n) * h(n)$$

Convolution in the time domain becomes a multiplication in the DTFT domain, hence we get

$$Y(\Omega) = X(\Omega) \cdot H(\Omega)$$

where $Y(\Omega), X(\Omega), H(\Omega)$ are the DTFT of $y(n), x(n), h(n)$ respectively. This means, if we want to see which signal frequencies Ω are attenuated and which are passed, we just need to know $H(\Omega)$, the filters "frequency response".

1.4 Sampling an Analog Signal

The audio signal is sampled from the analog microphone signal to a time-discrete version, sampled at a certain frequency, for instance at 44.1 kHz for an audio CD. Then the maximum audio frequency is less than 22.05 kHz, which is called the Nyquist frequency. After sampling, the audio samples are independent of the physical time, hence the frequencies in that domain are normalized such that 2π represents the sampling frequency and π the Nyquist frequency [6]. To see this we can simply express the sampling of an analog audio signal $s(t)$ as multiplying it with dirac pulses $\delta(nT - t)$ at the sampling time instances nT, with the sampling period T. The Fourier domain representation or spectrum of the continuous analog audio signal $s(t)$ is $S^c(\omega)$ (c for "continuous"),

$$S^c(\omega) = \int_{t=-\infty}^{\infty} s(t) \cdot e^{-j\omega t} dt$$

with $\omega = 2\pi f$ for frequency f. Because the integral over the signal multiplied with a dirac impulse is simply the signal value at the dirac position, the integral turns into a sum after sampling of the audio signal, and we get the spectrum $S^d(\omega)$ (with d for "discrete"),

$$S^d(\omega) = \sum_{n=-\infty}^{\infty} s(nT) \cdot e^{-j\omega nT} \tag{1.5}$$

Observe that the frequency expression in the sum only appears together with T, as $\omega \cdot T$. Now the inverse of the sampling period is the sampling frequency $f_s = 1/T$. Hence we can rewrite it as

$$\omega \cdot T = \omega / f_s =: \Omega$$

This is now exactly our normalized frequency, where 2π represents the sampling frequency f_s! To indicate that we are now in the discrete time domain, we rename the signal as

$$x(n) = s(nT).$$

Its frequency domain description then becomes

$$X(\Omega) = \sum_{n=-\infty}^{\infty} x(n) \cdot e^{-j\Omega n}$$

Since we now have integers as factors of Ω in the exponent of e, the spectrum $X(\Omega)$ is now periodic in Ω with a period of 2π. This is called the Discrete Time Fourier transform [6].

Also observe that for real valued signals, the spectrum for negative frequencies is the conjugate complex spectrum of the positive frequencies,

$$X(-\Omega) = X^*(\Omega)$$

(where $.^*$ is the conjugate complex operation), because $e^{-j(-\Omega)n} = \left(e^{-j\Omega n}\right)^*$.

1.5 Downsampling a Time-Discrete Signal

Our filter bank uses downsampling and upsampling for our already time discrete signal. So what happens when we then downsample the already time discrete signal $x(n)$, to reduce its sampling rate? Downsampling by N in discrete time means we only keep every Nth sample and discard every value in between. This can now be described as first multiplying our signal with a sequence of delta impulses (a 1 at each sample position), zeros in between [5], which we also call a delta train. Later we can drop the zeros in our signal. We use the delta train as defined in (1.1)

$$\delta_N(n) = \begin{cases} 1 & \text{if} \quad n = mN \\ 0 & \text{else} \end{cases}$$

We obtain the downsampled signal, still containing the zeros between the samples, as

$$x^d(n) = x(n) \cdot \delta_N(n)$$

The frequency domain representation of the signal multiplied with the delta train then becomes

$$X^d(\Omega) = \sum_{n=mN} x(n) \cdot e^{-j\Omega n}$$

$$= \sum_{m=-\infty}^{\infty} x(mN) \cdot e^{-j\Omega mN}$$

for all integers m. How is this spectrum of the downsampled signal related to the spectrum of the original signal? To see this, we apply a trick. We write the sequence of delta impulses as a sum of exponentials,

$$\delta_N(n) = \frac{1}{N} \sum_{k=0}^{N-1} e^{j\frac{2\pi}{N} \cdot k \cdot n} = \begin{cases} 1 & \text{if} \quad n = mN \\ 0 & \text{else} \end{cases} \tag{1.6}$$

If n is an integer multiple of N, as for $n = mN$, then this sum is 1 because $e^{j\frac{2\pi}{N} \cdot mN \cdot i} = e^{j2\pi \cdot m \cdot i} = 1$. If n is not a multiple of N, we see that this is a geometric sum (a sum over a constant with the summation index in the exponent). It can be written as $S := \sum_{k=0}^{N-1} c^k$, where we have $c = e^{j2\pi/N \cdot n}$. It is easy to see that $S \cdot c = \sum_{k=1}^{N} c^k$. Hence we get $S \cdot c - S = c^N - 1$, leading to

$$S = \frac{c^N - 1}{c - 1}$$

This means our sum has the closed form solution

$$\sum_{k=0}^{N-1} e^{j\frac{2\pi}{N} \cdot n \cdot k} = \frac{e^{j\frac{2\pi}{N} \cdot nN} - 1}{e^{j\frac{2\pi}{N} \cdot n} - 1} = 0$$

for n not a multiple of N, which proves what is stated in Eq. (1.6). Observe that this leads to sampling with phase 0.

We know that a multiplication of two sequences in the time domain leads to a convolution of their Fourier transforms in the frequency domain. Hence multiplying our audio signal $x(n)$ with the impulse train with phase shift n_0, $\delta_N(n - n_0)$ leads to convolving the frequency response $X(\Omega)$ with the frequency response of $\delta_N(n - n_0)$, the sequence of dirac impulses. This convolution is the sum of the original frequency response of $x(n)$, each shifted by the frequency location of one of the dirac impulses. This result is also obtained by directly multiplying Eq. (1.6) with $x(n)$ in the time domain,

$$x_{n_0}^d(n) = x(n) \cdot \delta_N(n - n_0) = x(n) \cdot \frac{1}{N} \sum_{k=0}^{N-1} e^{j\frac{2\pi}{N} \cdot k \cdot (n - n_0)}$$

now with the subscript n_0 to denote the phase, and taking the Discrete Time Fourier transform,

$$X_{n_0}^d(\Omega) = \sum_{n=-\infty}^{\infty} x(n) \cdot \frac{1}{N} \sum_{k=0}^{N-1} e^{j\frac{2\pi}{N} \cdot k \cdot (n - n_0)} \cdot e^{-j\Omega n}$$

$$= \frac{1}{N} \sum_{k=0}^{N-1} e^{j\frac{2\pi}{N} \cdot k \cdot n_0} \cdot \sum_{n=-\infty}^{\infty} x(n) \cdot e^{-j(-\frac{2\pi}{N} \cdot k + \Omega) \cdot n}$$

$$= \frac{1}{N} \sum_{k=0}^{N-1} e^{j\frac{2\pi}{N} \cdot k \cdot n_0} \cdot X\left(-\frac{2\pi}{N} \cdot k + \Omega\right)$$

or in short,

$$X_{n_0}^d(\Omega) = \frac{1}{N} \sum_{k=0}^{N-1} e^{j\frac{2\pi}{N} \cdot k \cdot n_0} \cdot X\left(-\frac{2\pi}{N} \cdot k + \Omega\right) \tag{1.7}$$

This final result now again shows the connection between the frequency response of the original signal and of the downsampled signal (still containing the zeros), that it is the sum of the original frequency response of $x(n)$, each shifted by the frequency location of one of the dirac impulses in the frequency domain. The frequency shifted versions of the original spectrum in this sum are called aliasing, aliasing components, or aliased frequencies.

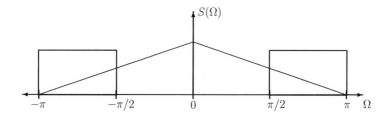

Figure 1.2: The magnitude spectrum of a signal. The two boxes symbolize the passband of the ideal bandpass. Here they are part of a 2-band filter bank.

Observe that this formulation is very similar to the case of sampling a continuous signal, except that the frequency responses or spectra are now periodic, and the sum with the aliasing components is finite. The spectrum of the original signal was 2π periodic, but observe that the signal multiplied with the delta impulse train is now **$2\pi/N$ periodic** because of the sum with the shifted spectra, the aliasing components. This means that, for instance, the frequency range from $-\pi/N$ to π/N contains a full and complete description of the spectrum of the signal.

1.5.1 Python Example Sampling with Unit Impulse Train

Python is an interpreted object-oriented programming language with increasing popularity also in the signal processing community. It is open source and available for most operating systems, most notably Linux. It has a convenient interactive front end (iPython), and a wide range of libraries, for instance "pylab" for signal processing and mathematical computations and plots.

Our **example** can be seen in Fig. 1.2. It shows the magnitude response of a bandpass, in this example a highpass filter as part of a 2-band filter bank, on top of the spectrum of a signal to be filtered.

1 Filter Banks

Figure 1.3 shows the resulting signal spectrum after passing this high pass. This signal is now multiplied by the delta impulse train. The resulting signal spectrum, still containing the zeros, can be seen in Fig. 1.4. Observe that the signal spectrum of the high pass signal appears mirrored around frequency zero, meaning the high and low frequencies within the subband switched their positions, high frequencies became low and vice versa. In general this does not happen to all subbands. Using the above principles it can be seen that this happens only to all the odd numbered subbands, if the lowest subband starts with the number 0.

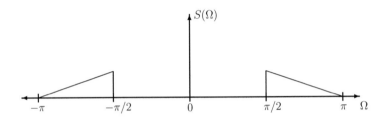

Figure 1.3: Signal spectrum after passing the bandpass.

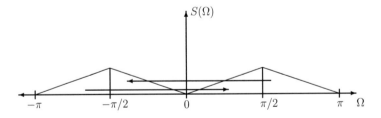

Figure 1.4: Signal spectrum after setting all samples to zero except every N-th (every second in this example). Observe that this step generates a copy of the negative band starting at $-\pi$ at the positive band starting at frequency 0, and a copy of the positive band below frequency $+\pi$ at the negative band below frequency 0, as indicated by the arrows.

This behaviour of flipping the frequencies of every second band can also be seen in the following iPython example. In a terminal we start iPython with library pylab with
`ipython --pylab`
To simulate a bandpass signal, we generate two sinusoids who are closely spaced together in frequency and with different amplitudes:
`sig = sin(pi*0.4*arange(1024))+0.5*sin(pi*0.35*arange(1024))`
Then we generate the frequency response on the whole unit circle, meaning including the negative frequencies, with the option "whole", and plot the result,

```
from scipy.signal import freqz
w, H = freqz(sig,1,512,'whole')
plot(w,20*log10(abs(H)))
xlabel('normalized frequency (pi is Nyquist frequency)');
ylabel('dB');
```

The resulting magnitude of the frequency response is shown in Fig. 1.5. Observe that the positive frequencies appear between normalized frequencies 0 and π, and the negative frequencies between π and 2π!

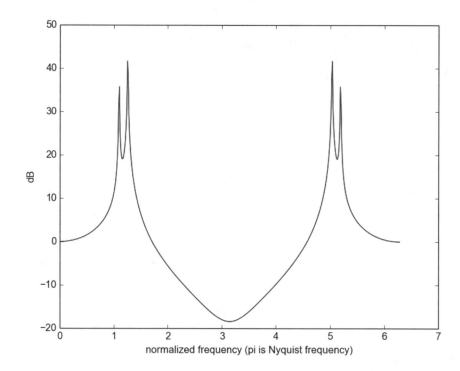

Figure 1.5: Magnitude of the frequency response of two sine waves.

Now we multiply it with our unit impulse train with $N = 4$, hence keeping every 4th sample and setting samples in between to zero,

```
dsig=zeros(size(sig));
dsig[0::4]=sig[0::4];
```

```
w, Hdsig=freqz(b=dsig,a=1,worN=512,'whole');
plot(w,20*log10(abs(Hdsig)))
xlabel('normalized frequency (pi is Nyquist frequency)')
ylabel('dB')
```

Figure 1.6 shows the resulting magnitude of the frequency response. Observe that our original positive and negative "subbands" are each shifted and copied by multiple frequencies of $2 \cdot \pi/4$, and hence in effect every second subband appears mirrored in its frequency content. Since in this plot the Nyquist frequency corresponds to 1 instead of π, we obtain frequency shifts of $2/4 = 0.5$.

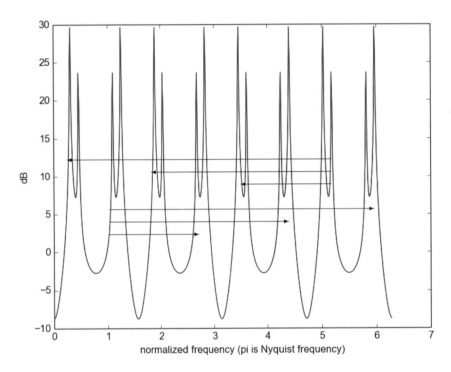

Figure 1.6: Magnitude of the frequency response of the two sine waves, multiplied with a unit pulse train with $N = 4$. Observe the copied and shifted spectral images indicated by the arrows.

1.5.2 Removal of the Zeros

The final step of downsampling is now to **omit the zeros between the samples**, to finally obtain the lower sampling rate. The downsampled signal, without the zeros between the samples, is $x_0^{\downarrow N}(m)$, with sampling phase 0, and where the time index m denotes the lower sampling rate, as opposed to n, which denotes sampling at the higher sampling rate,

$$x_0^{\downarrow N}(m) = x(mN + 0) = x_0^d(mN)$$

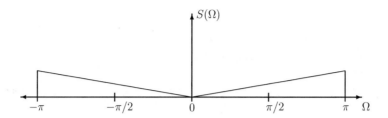

Figure 1.7: Signal spectrum after downsampling by N (2 in this example).

How are the frequency responses of $x_0^d(n)$ and $x_0^{\downarrow N}(m)$ connected? To answer this question, we can simply take the Fourier transforms of them,

$$X_0^d(\Omega) = \sum_{n=-\infty}^{\infty} x_0^d(n) \cdot e^{-j\Omega n}$$

$$= \sum_{n=mN} x_0^d(n) \cdot e^{-j\Omega n}$$

$$= \sum_{m=-\infty}^{\infty} x_0^{\downarrow N}(m) \cdot e^{-j\Omega mN}$$

$$= X_0^{\downarrow N}(\Omega N) \tag{1.8}$$

or, equivalently

$$X_0^{\downarrow N}(\Omega) = X_0^d(\Omega/N).$$

This means that all that is happening by omitting the zeros is a stretching of the frequency scale. For instance, the normalized frequency π/N before downsampling becomes π after removing the zeros. With this stretching of the spectrum we don't lose information in this case, because before removing the zeros we already had a complete spectrum of our signal in the narrower range of $-\pi/N$ to π/N. For our high pass example this can be seen in Fig. 1.7.

So all together, with (1.8) and (1.7) for downsampling by N with phase n_0, including the removal of the zeros between samples, we obtain

$$X_{n_0}^{\downarrow N}(\Omega) = \frac{1}{N} \sum_{k=0}^{N-1} e^{j\frac{2\pi}{N} \cdot k \cdot n_0} \cdot X\left(-\frac{2\pi}{N} \cdot k + \frac{\Omega}{N}\right) \tag{1.9}$$

1.5.3 Python Example

Make a sine wave which at 44,100 Hz sampling rate has a frequency of 400 Hz at 1 s duration. Hence we need 44,100 samples, and 400 periods of our sinusoid in this second. Hence we can write our signal in Python and listen to it with the following code:

```
import pyaudio
s=sin(2*pi*400*arange(0.,1.,1./44100))
#init audio stream to play array data
p = pyaudio.PyAudio()
stream = p.open(format=pyaudio.paFloat32,channels=1,rate=44100,
output=True)
stream.write(s.astype(float32))
```
Now plot the first 1000 samples:
```
plot(s[0:1000])
plt.xlabel('Samples')
ylabel('Amplitude'),
```
see Fig. 1.8. Next plot the first 100 samples:

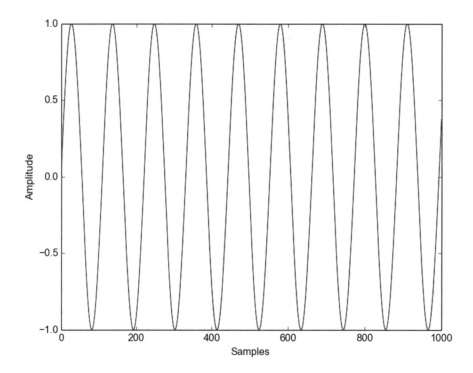

Figure 1.8: The first 1000 samples of our 400 Hz sine signal.

```
plot(s[0:100]),
```
see Fig. 1.9. We can take a look at the spectrum of the original signal s with:
```
from scipy.signal import freqz
w, H = freqz(s,1,worN=2048)
plot(w,20*log10(abs(H)))
ylim(-50,80)
xlim(0,3.14) xlabel('Normalized Frequency (π is Nyquist)')
ylabel('Magnitude')
```
see Fig. 1.10. Figure 1.10 shows the magnitude of the frequency spectrum of our signal. Observe that the frequency axis (horizontal) is a normalized frequency, where the

Nyquist frequency, in our case 22,050 Hz, appears as π. Hence our sinusoid should appear as a peak at normalized frequency $400/22050 \cdot \pi \approx 0.057$, which we indeed see.

Now we can multiply this sine tone signal with a unit pulse train, with N=8. We use an indexing trick to get the desired result of only keeping every 8th sample and having zeros in between, and then plot the first 100 samples::

```
sdu=zeros(size(s))
sdu[0::8]=s[0::8]
plot(sdu[1:100])
xlabel('Sample')
ylabel('Value')
```

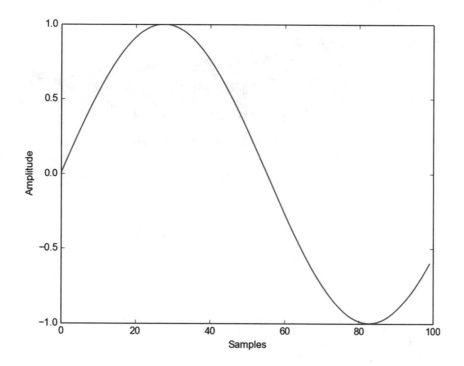

Figure 1.9: The first 100 samples of our 400 Hz sine signal.

see Fig. 1.11. Now we can compare the spectra of our original sine signal to our down-sampled sine signal with the zeros still in it:

```
from scipy.signal import freqz
w, h = freqz(sdu,worN=2048)
plot(w, 20*log10(abs(h)))
ylabel('dB') xlabel('Normalized frequency') ,
```

see Fig. 1.12. Here we can see the original line of our 400 Hz tone, and now also the 7 new aliasing components. Observe that always 2 aliasing components are close together. This is because the original 400 Hz tone also has a spectral peak at the negative frequencies, at -400 Hz, or at normalized frequency -0.018. This is also shifted by multiples of $2\pi/N$, just like the positive frequencies.

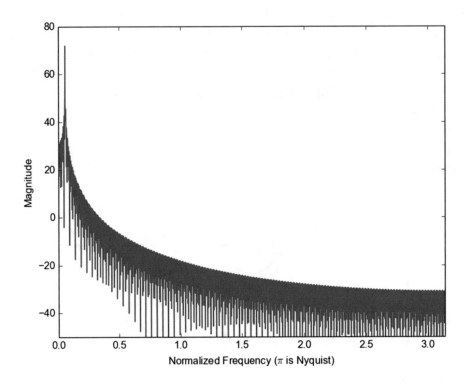

Figure 1.10: The spectrum of our 400 Hz sine signal.

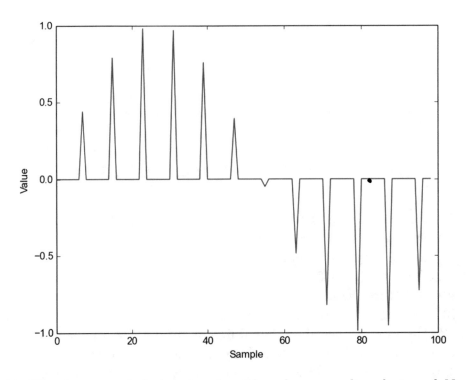

Figure 1.11: The downsampled sine signal, with a downsampling factor of $N = 8$, and with the zeros between the samples still in it.

Now also listen to the signal with the zeros:

```
stream.write(sdu.astype(float32))
```

Here you can hear that it sounds quite different from the original, because of the strong aliasing components!

The final step of downsampling is now to omit the zeros between the samples, to obtain the lower sampling rate. Let's call the signal without the zeros sd. In our iPython example this is:

```
sd=sdu[0:44100:8]
plot(sd[1:(100/8)])
xlabel('Sample')
ylabel('Value')
```

see Fig. 1.13. We can now take a look at the spectrum with

```
w, h = freqz(sd,worN=2048)
plot(w, 20*log10(abs(h)))
ylabel('dB')
xlabel('Normalized frequency')
```

see Fig. 1.14. Observe that the sine signal now appears at normalized frequency of 0.46, a factor of 8 higher than at the higher sampling rate. This is because we now have a new Nyquist frequency of 22050/8 now, hence our normalized frequency becomes $400/22050 \cdot \pi \cdot 8$. This shows that removing the zeros scales or stretches our frequency axis.

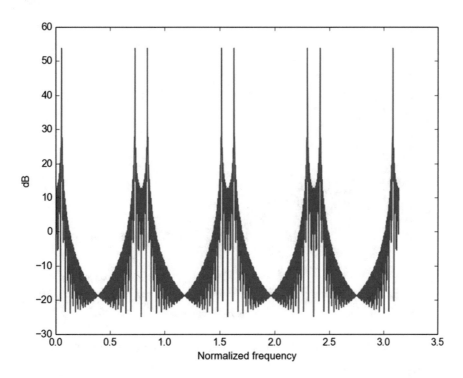

Figure 1.12: The spectrum of our downsampled sine signal, with the zeros still in it.

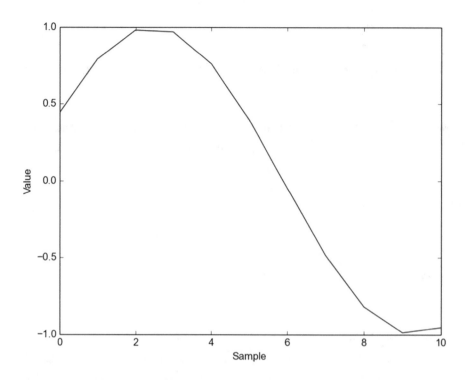

Figure 1.13: Our downsampled sine signal, without the zeros in it.

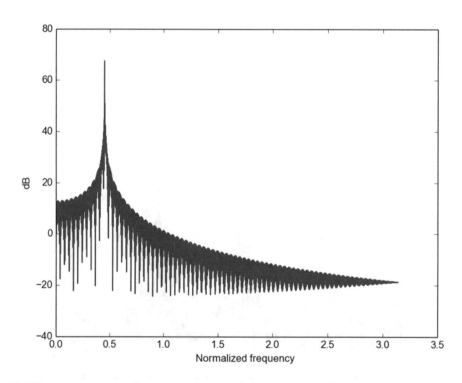

Figure 1.14: The spectrum of our downsampled sine signal, without the zeros between the samples.

1.6 Upsampling

We still need to analyse the opposite of downsampling, the upsampling, for the case if we want to increase the sampling rate of our audio signal. For instance, upsampling our k'th subband signal $y_k(m)$ by a factor of N means that we insert $N-1$ new samples after each original sample, such that we increase the number of samples by a factor of N. One of the first approaches that often come to mind is to simply repeat each sample $N-1$ times. But this is equivalent to first inserting $N-1$ zeros after each original sample and then low pass filtering this new sequence by a low pass filter whose impulse response or filter coefficients consist of N ones. To obtain a more general approach, we upsample our signal in general by first inserting $N-1$ zeros, and then have a general interpolation filter to filter this sequence [5].

To distinguish the time indices at the different sampling rates, we again take m for the lower sampling rate and n for the higher sampling rate.

Synthesis

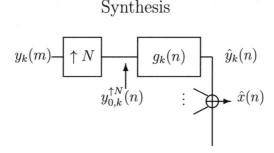

Figure 1.15: Upsampling and filtering in the synthesis filter bank for subband k and phase 0.

Figure 1.15 illustrates the upsampling as part of the synthesis filter bank, for subband k. Observe that this is simply the reverse step of the removal of the zeros as the last step of the downsampling. Hence we also get the connection between the frequency responses of the versions with the lower and the higher sampling rate in the same way as in Eq. (1.8),

$$Y_{0,k}^{\uparrow N}(\Omega) = Y_k(\Omega N). \qquad (1.10)$$

This means that increasing the sampling rate by a factor of N is a shrinking of the frequency scale. If our normalized frequency for Y was π, then after upsampling it is π/N. Since our spectra are 2π periodic, the spectrum of X^d becomes $2\pi/N$ periodic, which means we obtain aliasing components at frequency distances of $2\pi/N$, just as in Fig. 1.4. If we want to reconstruct the original high-band signal, we just need to apply our ideal high-pass filter again, such that we obtain the spectrum of Fig. 1.3. Observe that this seems to contradict the traditional Nyquist Theorem, because in this case our sampling rate is not twice the upper limit of our signal frequencies. The difference is that here we sample a bandpass signal, and our sampling rate is twice the **bandwidth** of our signal. This principle can also be reduced to the traditional Nyquist Theorem if we view the aliasing introduced by the sampling as moving the bandpass signal down to the baseband, around frequency zero, and then sampling it with the traditional Nyquist Theorem. Reconstruction is then obtained by using the aliasing from upsampling to shift the signal frequency at its original place in the spectrum, with filtering to suppress the unwanted aliased parts.

The reconstruction in the synthesis filter bank uses the upsampling, followed by the filtering for all subbands. Observe that the original subband signals can be reconstructed, because their spectrum does not overlap with any aliased spectral parts. To obtain the original signal, simply all the subband signals are added to result in the original spectrum and hence the original signal.

1.7 The z-Transform and Effects in the z-Domain

Since the z-Transform is a more general transform, which we will use to obtain filter banks with perfect reconstruction, it is helpful to look at the effects the sampling operations have in the z-domain. For practical purposes we usually only consider causal systems and signals, hence we use the 1-sided z-Transform, defined as [6]

$$X(z) = \sum_{n=0}^{\infty} x(n) z^{-n}$$

- The first useful observation is that we can obtain our usual **frequency response** if we evaluate the z-transform along the unit circle in the z-domain,

$$z = e^{-j\Omega}.$$

- One of the most important advantages of the z-transform is that it turns a **convolution** in the time domain **into a (mathematically more simple) multiplication in the z-domain**. This can be easily seen since the z-transform turns a sequence into a polynomial, and a convolution $x(n) * y(n)$ of two sequences $x(n), y(n)$ is mathematically identical to the multiplication of its associated polynomials,

$$a(n) * b(n) \rightarrow A(z) \cdot B(z)$$

- A further useful tool is the **reversal of the ordering** of a finite length sequence, with length L (meaning $x(n)$ is non-zero only for $n = 0, \ldots, L-1$),

$$x_r(n) := x(-n)$$

Its z-transform is

$$X_r(z) = \sum_{n=0}^{L-1} x(-n) z^{-n} =$$

with the index substitution of n by $-n$ this becomes

$$= \sum_{n=0}^{L-1} x(n) z^{-(-n)} =$$

$$= X(z^{-1})$$

or in short,

$$X_r(z) = X(z^{-1}).$$

This means that a reversal of the ordering of a sequence results in inverting z in its z-transform. If we take

$$x_r(n) := x(L - 1 - n)$$

we need to take into account the delay be $N - 1$ samples compared to simply replacing n by $-n$, which then results in

$$X_r(z) = z^{-L+1} \cdot X(z^{-1}).$$

- Remember that the convolution was defined as

$$x(n) * y(n) = \sum_{m=-\infty}^{\infty} x(m) \cdot y(n - m)$$

(1.4). This can now be used for the **correlation coefficients** $r_{xy}(n)$ of two sequences $x(n), y(n)$, defined as

$$r_{xy}(n) = \sum_{m=\infty}^{\infty} x(m) \cdot y^*(m - n)$$

This can also be written as the convolution of the signal $x(n)$ with the time-reversed signal $y(-n)$,

$$r_{xy}(n) = x(n) * y^*(-n)$$

similarly, the autocorrelation coefficients $r_{xx}(n)$ can be written as

$$r_{xx}(n) = x(n) * x^*(-n)$$

Since the convolution in the time domain becomes a multiplication in the z-domain, we get the z-domain expression

$$R_{xx}(z) = X(z) \cdot X^*(z^{-1}) \tag{1.11}$$

or in the DTFT domain (with the substitution $z = e^{j\omega}$)

$$R_{xx}(e^{j\omega}) = X(e^{j\omega}) \cdot X^*(e^{-j\omega})$$

Here we can see that in this case we have a product of a number with its conjugate complex version, hence the squared magnitude,

$$X(e^{j\omega}) \cdot X^*(e^{-j\omega}) = X(e^{j\omega}) \cdot \overline{X(e^{j\omega})} = |X(e^{j\omega})|^2 \tag{1.12}$$

This shows that the DTFT of the autocorrelation coefficients $r_{xx}(n)$ is the **power spectrum** of the signal $x(n)$ (the magnitude squared spectrum).

- What is the effect of **multiplying our signal with the delta impulse train** in the z-domain? We can see it by applying the z-transform to it, using our reformulation for the delta impulse train of Eq. (1.6),

$$X^d(z) = \sum_{n=0}^{\infty} x^d(n)z^{-n}$$

$$= \sum_{n=0}^{\infty} x(n) \cdot \delta_N(n)z^{-n}$$

$$= \frac{1}{N} \sum_{k=0}^{N-1} \sum_{n=0}^{\infty} x(n) \cdot \left(e^{-j\frac{2\pi}{N}\cdot k} \cdot z\right)^{-n}$$

$$= \frac{1}{N} \sum_{k=0}^{N-1} X\left(e^{-j\frac{2\pi}{N}\cdot k} \cdot z\right)$$

or in short,

$$X^d(z) = \frac{1}{N} \sum_{k=0}^{N-1} X\left(e^{-j\frac{2\pi}{N}\cdot k} \cdot z\right) \tag{1.13}$$

This means, in the z-domain the aliasing components appear with the exponential function multiplied with z.

- The next effect is the **removal of the zeros** from or into the signal. Let's again use our definition $x^{\downarrow N_0}(m) = x^d(mN)$. Then the z-transform becomes

$$X_0^{\downarrow N}(z) = \sum_{m=0}^{\infty} x^{\downarrow N}(m)z^{-m}$$

$$= \sum_{m=0}^{\infty} x^d(mN)z^{-m}$$

$$= \sum_{n=0}^{\infty} x^d(n)z^{-n/N} = X^d(z^{1/N})$$

or in short,

$$X^{\downarrow N}(z) = X^d(z^{1/N}).$$

- **Upsampling** by insertion of zeros is basically equivalent to the removal of the zeros above, but reverse. For the z-transform we get

$$Y^{\uparrow N}(z) = \sum_{n=0}^{\infty} y^{\uparrow N}(n)z^{-n} = \sum_{m=0}^{\infty} y(m)z^{-mN} = Y(z^N) \tag{1.14}$$

This means that upsampling by inserting $N-1$ zeros after each sample leads to a z-transform where we replace each z by z^N.

- Another useful tool is the so-called **modulation**, which is the multiplication of our signal with a modulation function, for instance an exponential function,

$$x_M(n) := x(n) \cdot e^{j\Omega_M \cdot n}.$$

Its z-transform becomes

$$X_M(z) = X(e^{j\Omega_M} \cdot z).$$

Observe that this has the effect of simply **shifting the frequency response** by Ω_M, which can be seen by simply replacing z by $e^{j\Omega}$ to obtain the frequency response. We obtain an interesting special case, if we set $\Omega_M = \pi$. In this case the modulation function is simply a sequence of ones with alternating signs, $1, -1, 1, -1, \ldots$ If we look at the resulting frequency response we see that the resulting frequency response after modulation is simply shifted by π. We know that real valued signals are symmetric around frequency 0, and that our sampled signal has a 2π periodic spectrum. Hence the resulting spectrum is obtained by simply **mirroring the spectrum** at frequency π. In this way, we can turn a **low pass into a high pass**, and vice versa.

1.8 Non-Ideal Filters

Our goal is to obtain perfect reconstruction after the synthesis filter bank with non-ideal filters. What happens when we use non-ideal filters instead of ideal filters? In the analysis filter bank also frequencies outside our desired pass band would get through, although at a reduced level. This is depicted in Fig. 1.16, which is the accordingly modified version of Fig. 1.3. In the figure, frequencies slightly lower than $\pi/2$, outside our desired passband, are also passed through. After downsampling, these frequencies outside our desired pass band would be aliased into our passband. This can also be seen in Fig. 1.17 in comparison to Fig. 1.4 with the ideal bandpass filters. These signal parts are also called aliasing.

In the synthesis filter bank, non-ideal filters mean that not only the desired pass band is generated, but also neighbouring frequencies, but at a lower level. That means that neighbouring bands have frequency components that are identical, which are added at the output of the synthesis filter bank. If we are lucky, the aliasing components cancel when adding up all the subbands in the synthesis filter bank at this point. That means, if we want to design realizable filter banks with perfect reconstruction, we have to design the filters this way. But observe that even though we can obtain perfect reconstruction after the synthesis filter bank, we still have aliasing in the subbands after analysis filtering and downsampling. This fact becomes important, for instance when a single band is omitted. In that case the alias cancellation with this band does not take place anymore, and it will show up at the reconstructed output. This alias cancellation has the biggest effect towards the neighbouring bands, because there the aliasing will have the highest magnitudes. But in general it will also affect more distant subbands, although at a lower magnitude.

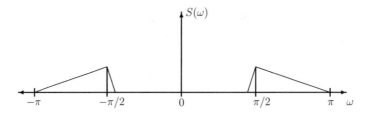

Figure 1.16: Signal spectrum after passing the non-ideal bandpass.

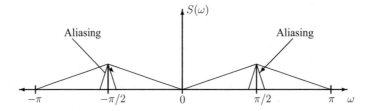

Figure 1.17: Signal spectrum after setting all samples to zero except every N-th (every second in this example), with the non-ideal bandpass. Observe the partially overlapping spectral parts, which results in aliasing in the downsampled subband signals.

We would like to have a filter bank which is switchable to different numbers of subbands during processing the audio signal, while maintaining perfect reconstruction. For an FFT it is straightforward to switch the number of subbands during processing. All that needs to be done is to change the size of the block of audio samples accordingly during processing. It becomes more complicated when filters are involved which are longer than the blocksize, and hence overlap to other blocks. In this case a transition between the different numbers of subbands has to be found in such a way that alias cancellation after synthesis filter bank is still provided during the transition.

1.9 Filter Bank Design for Perfect Reconstruction

To obtain a filter bank design method which guarantees us perfect reconstruction, a suitable mathematical description is important. This mathematical description is the use of z-Transforms and of the so-called polyphase matrices. Let's start with a short review of z-Transforms. More details can be found, for instance, in [6]. For a given time-sequence $x(n)$, where n is the sample index (an integer number), the (1-sided) z-Transform is defined as

$$X(z) = \sum_{n=0}^{\infty} x(n)z^{-n}$$

The property which becomes most important in our context is that the z-Transform turns a convolution of time sequences into a multiplication of their z-Transforms. More specifically, if

$$c(n) = a(n) * b(n)$$

then

$$C(z) = A(z) \cdot B(z)$$

This means that the convolution is turned into a polynomial multiplication. This can also be easily visualized by comparing the equations for the convolution and for the resulting coefficients after the polynomial multiplication. Recall that a convolution is defined by

$$a(n) * b(n) := \sum_{i=-\infty}^{\infty} a(n-i)b(i) = \sum_{i=-\infty}^{\infty} a(i)b(n-i)$$

if $a(n)$ and $b(n)$ are causal sequences, meaning that they have only zero entries for $n < 0$. One can ask "why is a multiplication more beneficial than a convolution?". The answer is that there are more mathematical tools available for the inversion of multiplications, unlike the case of the convolution, which will be important for the inversion of the analysis filter bank operation.

1.9.1 Analysis Filter Bank

In our filter bank we not only have convolutions, but also downsamplers and upsamplers. Particularly, the output of the k-th analysis filter after downsampling by N is obtained by the substitution $n = mN + N - 1$ (compare to Fig. 1.1). Here, the "$N - 1$" is the phase of the downsampled subband signal. We choose this phase because this is the usual implementation in coders (at the end of a "block" of N samples).

$$y_k(m) = y_{N-1,k}^{\downarrow N}(m) = \sum_{n=0}^{L-1} x(mN + N - 1 - n)h_k(n), \qquad (1.15)$$

where L is the length of our filters h_k. We assume that L is a multiple of N, which can always be obtained by appending zeros, if necessary.

1.9.2 Synthesis Filter Bank

In our synthesis filter bank we first have upsamplers to the original sampling rate (obtained by inserting zeros into our subband signals), followed by the synthesis filters. For the syntheses upsampling we choose the phase to be 0. The synthesis filters can be seen as interpolation filters for each subband, since the upsampled subband signals contain $N - 1$ zero samples between each non-zero sample, and the synthesis filters "fill-in" or interpolate those gaps for a suitable centre frequency for each subband. Looking at

Fig. 1.15 the output of the k'th synthesis filter $g_k(n)$ (the filter impulse response is again assumed to have length L) can hence be written as

$$\hat{y}_k(n) = y_k^d(n) * g_k(n) = \sum_{n'=0}^{L-1} y_k^d(n-n') \cdot g_k(n').$$

Since $y_k^d(n-n')$ is only non- zero at multiples of N, we can replace the index n' by $m'N + n'$ and n by $mN + n$, to be able to address only those samples which are non-zero. This also leads to a block-wise processing.

$$\hat{y}_k(mN + n) = \sum_{m'=0}^{L/N-1} \sum_{n'=0}^{N-1} y_k^d(mN + n - m'N - n') g_k(m'N + n')$$

Here we can see that we obtain the non-zero samples of y_k^d with $n = n'$. We can then also replace the upsampled subbands $y_k^d(n)$ by the subbands at the lower sampling rate, $y_k(m)$, which leads to

$$\hat{y}_k(mN + n) = \sum_{m'=0}^{L/N-1} y_k(m - m') g_k(m'N + n). \tag{1.16}$$

The reconstructed signal $\hat{x}(n)$ is the sum of all the subband components,

$$\hat{x}(mN + n) = \sum_{k=0}^{N-1} \hat{y}_k(mN + n) = \sum_{k=0}^{N-1} \sum_{m'=0}^{L/N-1} y_k(m - m') g_k(m'N + n) \tag{1.17}$$

Observe that if the length of g_k is short, then only a few summands of (1.16) are non-zero. For instance, if $L = N$, then only 1 summand is non-zero; or if $L = 2N$, then 2 summands are non-zero. These two examples are indeed important special cases, as we will see.

How can we find a concise mathematical description, which includes the downsampling? To find that out we first take a detour to transforms and efficient implementations.

1.9.3 Block Transforms

The next important property after perfect reconstruction was an efficient implementation (meaning low computational complexity). A straightforward implementation of a filter bank as in Fig. 1.1 would lead to the following complexity. For filters of length L we have L multiplications per output sample. We have N filters, so for each new output sample we need NL multiplications. Fortunately we have the downsamplers after the analysis filters. This reduces the required computational complexity since we only need to compute the output samples which are kept by the downsampler, hence every N-th sample. In other words, we only need to compute the output of the filters once per block of N input samples. Hence we need NL multiplications per block of N input samples to obtain N subband samples (1 sample for each subband).

A well-known approach to obtain a computationally efficient subband decomposition is the Fast Fourier Transform (FFT) [6]. Per block of N input samples it only needs on the order of $N \log(N)$ multiplications to compute N subband samples, instead of NL. Since the filters of our filter bank usually have a length $L \geq N$, it can be seen that the FFT is computationally more efficient. There is another advantage: when we process the audio signal in adjacent non-overlapping blocks of length N, we not only have the critical sampling (since for every N input samples we obtain 1 subband sample for each of the N subbands), we also obtain perfect reconstruction with a synthesis filter bank which consists of simply applying the inverse FFT to each transformed block, and then reassembling the blocks into an audio stream.

1.10 Connection of Block Transforms to Filter Banks

1.10.1 DFT Analysis Transform

But what is the connection of a transform like a DFT or its fast implementation with an FFT to a filter bank? What kind of equivalent filters would an FFT have? [9].

For each input block we obtain N subband values, which are the $y_k(m)$, for $k = 0, \ldots, N-1$. We can write the FFT operation as a matrix multiplication with our input signal block. Since for the transform we use non-overlapping and adjacent blocks of the signal, we can write the m'th signal block as the row vector

$$\mathbf{x}(m) := [x(mN), x(mN+1), \ldots, x(mN+N-1)] \qquad (1.18)$$
$$= [x_0^{\downarrow N}(m), \ldots, x_{N-1}^{\downarrow N}(m)]$$

which consists of N input samples, and the rightmost element is the most recent sample. The DFT can be written as a multiplication with a transform matrix $\mathbf{T}$, whose entries consist of the Fourier transform kernel,

$$\mathbf{T}_{n,k} = e^{-j\frac{2\pi}{N} \cdot n \cdot k}$$

Similarly we define a row vector which contains the subband values (1 sample for each subband) for the m'th block as

$$\mathbf{y}(m) := [y_0(m), y_1(m), \ldots, y_{N-1}(m)]. \qquad (1.19)$$

Now we can write the DFT of the m'th block as

$$\mathbf{y}(m) = \mathbf{x}(m) \cdot \mathbf{T}$$

Observe that this matrix formulation easily leads to perfect reconstruction (using the inverse transform), even though we don't use ideal filters. For an extension to more general filter banks it will be useful to look at this transform formulation in the z-domain. Here we have sequences of vectors (the $\mathbf{y}(m)$ and $\mathbf{x}(m)$) instead of scalars, but we can still apply the z-transform in the same way,

$$\mathbf{Y}(z) = \sum_{m=0}^{\infty} \mathbf{x}(m) \cdot \mathbf{T} \cdot z^{-m} = \mathbf{X}(z) \cdot \mathbf{T}$$

meaning that the z-transform is now a sum of vectors multiplied with powers of z, and that the transform is still a simple matrix multiplication in the z-domain.

Now we can also extract the equivalent impulse responses of our transform. Looking at the k'th subband we get

$$y_k(m) = \sum_{n=0}^{N-1} x(mN+n) \cdot e^{-j\frac{2\pi}{N}\cdot n\cdot k} = \tag{1.20}$$

$$= \sum_{n=0}^{N-1} x(mN+N-1-n) \cdot e^{-j\frac{2\pi}{N}\cdot(N-1-n)\cdot k}$$

where we use a reversal of the summation order with $n \to N-1-n$. If we compare Eq. (1.20) with Eq. (1.15), we can see that the equivalent impulse response of our transform is the time reversed k'th column of the transform matrix,

$$h_k(n) = e^{-j\frac{2\pi}{N}\cdot(N-1-n)\cdot k}$$

This is now the equivalent analysis impulse response of our DFT, interpreted as analysis filter bank! So the main result here is that **each column of our transform matrix T represents the impulse response of one subband filter, but in reversed order.**

1.10.2 DFT Synthesis Transform

Similarly, for the synthesis we get for each block [9]

$$\hat{\mathbf{x}}(m) = \mathbf{x}(m) = \mathbf{y}(m) \cdot \mathbf{T}^{-1}$$

or, for the individual samples, using the inverse DFT transform matrix,

$$\hat{x}(mN+n) = \sum_{k=0}^{N-1} y_k(m) \cdot \frac{1}{N} \cdot e^{j\frac{2\pi}{N}\cdot n\cdot k}$$

Comparing it with Eq. (1.17) we can see that $L = N$ because the sum over m' disappeared (meaning $m' = 0$), and that the equivalent impulse response of the k'th synthesis filter is

$$g_k(n) = \frac{1}{N} \cdot e^{j\frac{2\pi}{N}\cdot n\cdot k}$$

for $n = 0,\ldots,N-1$. **The equivalent impulse responses of the synthesis transform are now identical to the rows of the transform matrix,** without time-reversal.

1.10.3 Discrete Cosine Transform

Another well-known example of a block transform is the Discrete Cosine Transform (DCT) [9, 10]. The so-called DCT Type 4 (short: DCT4) has the following analysis transform definition:

$$y_k(m) = \sqrt{\frac{2}{N}} \cdot \sum_{n=0}^{N-1} x(mN+n) \cdot \cos\left(\frac{\pi}{N}(k+0.5)(n+0.5)\right) \qquad (1.21)$$

Hence it has the analysis transform matrix

$$\mathbf{T}_{n,k} = \sqrt{\frac{2}{N}} \cdot \cos\left(\frac{\pi}{N}(k+0.5)(n+0.5)\right) \qquad (1.22)$$

Like above, the impulse responses of the equivalent filter bank are the time-reversed columns of the transform matrix,

$$h_k(n) = \sqrt{\frac{2}{N}} \cdot \cos\left(\frac{\pi}{N}(k+0.5)(N-1-n+0.5)\right)$$

The synthesis transform matrix is the inverse of the transform matrix, which is orthogonal. Hence its inverse is the transpose matrix, $\mathbf{T}^{-1} = \mathbf{T}^T$. The synthesis impulse responses are the rows of it,

$$g_k(n) = \sqrt{\frac{2}{N}} \cdot \cos\left(\frac{\pi}{N}(k+0.5)(n+0.5)\right)$$

One obvious property of the equivalent transform impulse responses is that their length is fixed to N, and hence do not overlap with neighbouring blocks. This property becomes a problem, for instance, when the subband values are quantized at different step sizes for consecutive blocks. This leads to audible discontinuities in the reconstructed audio signal at the block boundaries. In order to avoid those audible discontinuities we need longer filters which overlap with the neighbouring blocks. If we analyse the FFT filters we also see that they can be seen as the result of a multiplication of a rectangular window function with an infinite length exponential function, to obtain a finite length version. There is no tapering at the ends for a slow fade in or fade out. The multiplication with this rectangular window has the effect of convolving its frequency response with the frequency response of the exponential (a Dirac impulse). The frequency response of a rectangular window is a sinc function, which has relatively high sidelobes, which leads to only a relatively weak attenuation in the stop bands, which leads to more aliasing than necessary.

1.10.4 Short Time Fourier Transform (STFT)

An approach to introduce longer filter into the transform setting is using overlapping windows for the signal, and apply the DFT to them. This is the STFT [11–14]. For that, the signal is first divided into overlapping blocks of length N and with "hop-size" M. The analysis equation is

$$Y_k(m) = \sum_{n=0}^{N-1} h(n) \cdot x(m \cdot M + n)e^{-j\frac{2\pi}{N} \cdot k \cdot n}$$

with $m = 0$ until the end of the signal x the block index, and $h(n)$ is a window function of length N for improved filtering properties. We assume $N = LM$.

For the synthesis the overlapped blocks are added up (overlap-add). The synthesis equation is

$$\hat{x}_m(n) = \frac{h(n)}{N} \cdot \sum_{k=0}^{N-1} Y_k(m)e^{j\frac{2\pi}{N} \cdot k \cdot n}, n = 0, \ldots, N - 1$$

The overlap-add procedure for a block m_0 is:

$$x(m_0 \cdot M + n) = \sum_{m=0}^{L-1} \hat{x}_{m_0-m}(mM + n)$$

with $n = 0, \ldots, M - 1$. The window has the overlap-add property

$$\sum_{m=0}^{L-1} h^2(n + mM) = 1, n = 0, \ldots, M - 1$$

The STFT is also a filter bank, but with non-critical sampling, since usually $M < N$.

In the literature the window function $h(n)$ is usually only applied to the analysis part. Here we also applied it to the synthesis part, because this improves the resulting synthesis filters, and is more similar to usual filter banks. The corresponding Python function is `scipy.signal.stft`.

Now we have an idea how to obtain perfect reconstruction and an efficient implementation, using fast transforms like the FFT or the STFT. But the equivalent filters are either not good enough, because the filters are too short and have not enough stop band attenuation, or we have no critical sampling. Now we just have to find out how to extend this approach to more general filter banks, with longer and overlapping filters and critical sampling (with $M = N$). For that we come back to the mathematical description of filter banks.

1.10.5 Block Processing for Longer Filters

Analysis Filter Bank

Since after downsampling by N only every Nth sample is output, our approach is to view the processing of our time sequence as block-wise, with blocks of size N, just like we did with the transforms. We again use our block index m, and the index n for the position ("phase") inside this block.

We can now start rewriting (1.15), using the block indices m and m', and the phase index n, we apply the index substitution $n \Rightarrow m' + N - 1 - n$. Because we now have 2 indices, this results in the following double sum:

$$y_k(m) = y_{N-1,k}^{\downarrow N}(m) = \sum_{m'=0}^{L/N-1} \sum_{n=0}^{N-1} x(mN - m'N + n)h_k(m'N + N - 1 - n) \quad (1.23)$$

It can now be observed that the inner sum over the phase index n corresponds to the vector product of the vector $\mathbf{x}(m)$, which we already introduced for the transform implementation in Eq. (1.18), and a new vector which replaces the transform coefficients. This vector, which now contains the samples of our k'th filter impulse response, is also written as a sequence of blocks. It also uses a block index m, because the impulse response can be longer than one block,

$$\mathbf{h}_k(m) := [h_k(mN + N - 1), h_k(mN + N - 2), \ldots, h_k(mN)] \quad (1.24)$$

$$= [h_{N-1,k}^{\downarrow N}(m), h_{N-2,k}^{\downarrow N}(m), \ldots, h_{0,k}^{\downarrow N}(m)] \quad (1.25)$$

With the vector $\mathbf{h}_k(m)$ and with $\mathbf{x}(m)$ as defined in Eq. (1.18) the inner sum of Eq. (1.23) can be replaced by a vector multiplication,

$$y_k(m) = \sum_{m'=0}^{L/N-1} \mathbf{x}(m - m')\mathbf{h_k}^T(m')$$

This is now again in the form of a convolution, but now with sequences of vectors instead of samples,

$$y_k(m) = \mathbf{x}(m) * \mathbf{h}_k^T(m). \quad (1.26)$$

Hence we reached our goal to reduce the more complex form of filtering and downsampling to the simpler form of a plain convolution, where we can apply all the mathematical tools for a convolution. Using the z-transform, this expression can be mathematically further simplified by converting the convolution into a multiplication. The z-transform of the sequence of signal vectors $\mathbf{x}(m)$ of Eq. (1.18) can be obtained by applying the z-transform to the sequence of each of the in-block or the so-called phase components individually (because there is an array of phase components we call them "polyphase components", see also [4, 5, 15])

$$\mathbf{X}(z) = \sum_{m=0}^{\infty} \mathbf{x}(m)z^{-m} = [X_0^{\downarrow N}(z), X_1^{\downarrow N}(z), \ldots, X_{N-1}^{\downarrow N}(z)]$$

with the polyphase components

$$X_n^{\downarrow N}(z) = \sum_{m=0}^{\infty} x(mN + n)z^{-m} = \sum_{m=0}^{\infty} x_n^{\downarrow N}(m)z^{-m}.$$

1 Filter Banks

Similarly, the vector of the impulse response $\mathbf{h}_k(m)$ of Eq. (1.24) becomes

$$\mathbf{H}_k(z) := [H_{N-1,k}^{\downarrow N}(z), H_{N-2,k}^{\downarrow N}(z)), \ldots, H_{0,k}^{\downarrow N}(z)]$$

where each $H_{n,k}^{\downarrow N}(z)$ is the z-transform of $h_{n,k}^{\downarrow N}(m)$,

$$H_{n,k}^{\downarrow N}(z) = \sum_{m=0}^{\infty} h_k(mN + n)z^{-m}. \tag{1.27}$$

Note that we can obtain the z-transform $X(z)$ of the original signal $x(n)$ from its polyphase components with

$$X(z) = \sum_{n=0}^{N-1} X_n^{\downarrow N}(z^N) \cdot z^{-n} \tag{1.28}$$

The z-transform of the downsampled output of the k-th filter or our filter bank ($y_k(m)$ in Fig. 1.1) is obtained with

$$Y_k(z) = \sum_{m=0}^{\infty} y_k(m)z^{-m}.$$

For a single filter we don't obtain a vector.

We can now write the convolution in (1.26) as a multiplication of two vectors in the z-domain,

$$Y_k(z) = \mathbf{X}(z) \cdot \mathbf{H}_k^T(z) = \sum_{n=0}^{N-1} X_n^{\downarrow N}(z) \cdot H_{(N-1-n),k}^{\downarrow N}(z). \tag{1.29}$$

Synthesis Filter Bank

The synthesis equation can now also be reformulated to obtain a polyphase description of the synthesis filter bank. For Eq. (1.17) we already rewrote the synthesis equation for block-wise processing, to include the upsampling of the synthesis filter bank,

$$\hat{x}(mN + n) = \sum_{k=0}^{N-1} \sum_{m'=0}^{L/N-1} y_k(m - m')g_k(m'N + n)$$

Just like for the analysis filter bank, we define sequences of blocks as vectors for the reconstructed signal $\hat{\mathbf{x}}(m)$ and the synthesis impulse responses $\mathbf{g}_k(m)$, but unlike for the analysis filter bank, the vector for the synthesis filters now also has an increasing index order,

$$\hat{\mathbf{x}}(m) = [\hat{x}_0^{\downarrow N}(m), \hat{x}_1^{\downarrow N}(m), \ldots, \hat{x}_{N-1}^{\downarrow N}(m)] \tag{1.30}$$

$$\mathbf{g}_k(m) = [g_{k,0}^{\downarrow N}(m), g_{k,1}^{\downarrow N}(m), \ldots, g_{k,(N-1)}^{\downarrow N}(m)]. \tag{1.31}$$

Using these vectors we can simplify Eq. (1.17) to

$$\hat{\mathbf{x}}(m) = \sum_{k=0}^{N-1} \sum_{m'=0}^{L/N-1} y_k(m - m') \mathbf{g}_k(m')$$

Here we can see that the inner sum is in the form of a convolution. Hence we can turn it into a multiplication in the z-domain,

$$\hat{\mathbf{X}}(z) = \sum_{k=0}^{N-1} Y_k(z) \mathbf{G}_k(z)$$

This remaining sum is the sum over the N subbands, and can again be written as a vector multiplication with the vector $\mathbf{Y}(z)$, which contains the subband signals, and a synthesis polyphase matrix $\mathbf{G}(z)$,

$$\hat{\mathbf{X}}(z) = \mathbf{Y}(z)\mathbf{G}(z) \tag{1.32}$$

Visualization of Polyphase Elements and Blocks

To visualize the meaning of a polyphase vector, we take a signal $x(n)$ and its sequence of blocks, $\mathbf{x}(m)$, with m the block index, and a downsampling rate of $N = 2$. With this we get the following blocks and polyphase elements:

$$
\begin{array}{cc}
x_0^{\downarrow 2}(m) & x_1^{\downarrow 2}(m) \\
\text{phase0} & \text{phase1} \\
\downarrow & \downarrow \\
\end{array}
$$

$$
\begin{array}{ccc}
\mathbf{x}(0), \text{block0} \rightarrow & [x(0) & x(1)] \\
\mathbf{x}(1), \text{block1} \rightarrow & [x(2) & x(3)] \\
\end{array}
$$

Example for a Polyphase Vector We assume the block length and hence the number of subbands as $N = 8$, and our signal as the counting sequence with a length of 2 blocks,

$$x = [1, 2, 3, \ldots, 16]$$

then we obtain the time-domain polyphase vector as

$$\mathbf{x}(0) = [1, 2, \ldots, 8]$$

for block number $m = 0$, and

$$\mathbf{x}(1) = [9, 10, \ldots, 16]$$

for block number $m = 1$. Hence the z-domain polyphase vector is

$$\mathbf{X}(z) = [1 + 9 \cdot z^{-1}, 2 + 10 \cdot z^{-1}, \ldots, 8 + 16 \cdot z^{-1}]$$

Observe that we now have mathematically very simple operations for the analysis and synthesis filter bank, including downsampling and upsampling. This enables us to design filter banks with perfect reconstruction even though they have critical sampling and non-ideal filters. This formulation also yields another advantage. We can use it to shift the downsamplers of the analysis filter bank before the filtering or more general signal processing. This is in contrast to the original structure, where the filtering came first and then the downsampling. Analog for the synthesis filter bank, we can shift the upsamplers to after the filtering or rather signal processing. This is an advantage because all the signal processing takes place at the lower sampling rate, and hence reduces the computational complexity. The structure with the moved down- and upsamplers can be seen in Fig. 1.18.

1.11 Polyphase Matrices

For our filter bank we not only have one filter but we have N filters, $0 \leq k < N$, with N outputs $y_k(n)$. To account for these N filters, we arrange the N outputs of the filters also into a vector,

$$\mathbf{Y}(z) := [Y_0(z), Y_1(z), \ldots, Y_{N-1}(z)]$$

To obtain all the N outputs of the filter bank at once, we need to define N vectors of impulse responses $\mathbf{H}_k(z)$

$$\mathbf{H}_k(z) := [H_{N-1,k}^{\downarrow N}(z), H_{N-2,k}^{\downarrow N}(z)), \ldots, H_{0,k}^{\downarrow N}(z)],$$

where each $H_{n,k}^{\downarrow N}(z)$ is the z-transform of $h_{n,k}^{\downarrow N}(m)$. We assemble these vectors into a matrix $\mathbf{H}(z)$, where each column contains these polyphase components of one filter,

$$\mathbf{H}(z) := \begin{bmatrix} \mathbf{H}_0^T(z) & | & \mathbf{H}_1^T(z) & | & \cdots & | & \mathbf{H}_{N-1}^T(z) \end{bmatrix} =$$

$$= \begin{bmatrix} H_{N-1,0}^{\downarrow N}(z) & H_{N-1,1}^{\downarrow N}(z) & \cdots & H_{N-1,N-1}^{\downarrow N}(z) \\ H_{N-2,0}^{\downarrow N}(z) & & & \vdots \\ \vdots & & \ddots & \\ & \vdots & \vdots & \\ H_{0,0}^{\downarrow N}(z) & & \cdots & H_{0,N-1}^{\downarrow N}(z) \end{bmatrix} \tag{1.33}$$

with $H_{n,k}^{\downarrow N}(z)$ as shown in Eq. (1.27), as

$$H_{n,k}^{\downarrow N}(z) = \sum_{m=0}^{\infty} h_k(mN + n)z^{-m}. \tag{1.34}$$

This matrix is the polyphase matrix of the analysis filter bank (a polyphase matrix of type 1 [4]). With this matrix we now obtain a very simple mathematical formulation for the analysis filter bank, including the downsampling, which is simply a matrix multiplication (although with polynomials as matrix and vector elements),

$$\mathbf{Y}(z) = \mathbf{X}(z) \cdot \mathbf{H}(z) \tag{1.35}$$

Similarly, for the synthesis filter bank we have N filters, which we assemble in our synthesis polyphase matrix, as in (1.32)

$$\mathbf{G}(z) := \begin{bmatrix} \mathbf{G}_0(z) \\ \mathbf{G}_1(z) \\ \vdots \\ \mathbf{G}_{N-1}(z) \end{bmatrix} = \begin{bmatrix} G_{0,0}^{\downarrow N}(z) & G_{0,1}^{\downarrow N}(z) & \cdots & G_{0,N-1}^{\downarrow N}(z) \\ G_{1,0}^{\downarrow N}(z) & & & \vdots \\ \vdots & & \ddots & \\ & \vdots & \vdots & \\ G_{N-1,0}^{\downarrow N}(z) & & \cdots & G_{N-1,N-1}^{\downarrow N}(z) \end{bmatrix} \quad (1.36)$$

with

$$G_{k,n}^{\downarrow N}(z) = \sum_{m=0}^{\infty} g_k(mN+n)z^{-m}. \quad (1.37)$$

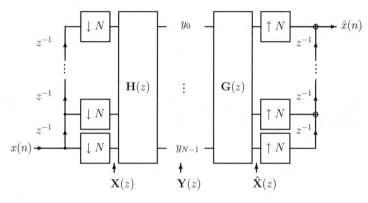

Figure 1.18: Polyphase representation of an N band filter bank with critical sampling. Observe that z^{-1} always means a delay by 1 sample, independent of the sampling rate.

This is the synthesis polyphase matrix. Observe that for this matrix, the indices for the subband k and the phase n are in switched order, compared to the analysis. For that reason this type is called polyphase matrix of type 2.

This reformulation now leads to a formulation of the analysis and syntheses filter banks as a mathematically much simpler matrix multiplication. On top of this advantage, it enables us to conduct all the signal processing at the lower sampling rate, and hence reduces the computational complexity. The structure with the moved down- and upsamplers can be seen in Fig. 1.18.

1.11.1 Converting Polyphase Representations

We can convert a polyphase representation at downsampling rate, for instance $2N$ and $X_n^{\downarrow 2N}(z)$, into a polyphase representation $X_n^{\downarrow N}(z)$ at downsampling rate N, as long as the higher sampling rate is an integer multiple of the lower sampling rate in general. In this example, we obtain the polyphase component at phase n at the lower sampling rate with

$$X_n^{\downarrow N}(z) = X_n^{\downarrow 2N}(z^2) + z^{-1} X_{N+n}^{\downarrow 2N}(z^2) \quad (1.38)$$

For the case of going back to the non-polyphase representation, this becomes

$$X(z) = \sum_{n=0}^{N-1} X_n^{\downarrow N}(z^N) \cdot z^{-n} \qquad (1.39)$$

1.11.2 Time-Reversal in Polyphase Components

We have a finite signal or impulse response $x(n)$ with length L, starting from 0. Its polyphase components are

$$X_n(z) = \sum_{m=-\infty}^{\infty} x(mN + n)z^{-m}$$

Now assume we replace $x(n)$ by $xr(n)$ with

$$xr(n) = x(L - 1 - n)$$

and assume that $L = l \cdot N$, which can always be made true, if necessary by appending zeros to $x(n)$. Its polyphase components are then

$$XR_n(z) = \sum_{m=-\infty}^{\infty} xr(mN + n)z^{-m}$$

$$= \sum_{m=-\infty}^{\infty} x(lN - 1 - mN - n)z^{-m}$$

if we replace n by $N - 1 - n$ and m by $-m$ we get

$$XR_{N-1-n}(z) = \sum_{m=-\infty}^{\infty} x(lN - 1 + mN - N + 1 + n)(z^{-1})^{-m}$$

$$XR_{N-1-n}(z) = \sum_{m=-\infty}^{\infty} x(N \cdot (l + m - 1) + n)(z^{-1})^{-m}$$

if we now replace m by $m - l + 1$ we get

$$XR_{N-1-n}(z) = \sum_{m=-\infty}^{\infty} x(N \cdot m + n)(z^{-1})^{-m} \cdot z^{-l+1} = z^{-l+1} \cdot X_n(z^{-1})$$

Hence all together we get

$$XR_n(z) = z^{-l+1} \cdot X_{N-1-n}(z^{-1}) \qquad (1.40)$$

1.11.3 Polyphase Representation for Correlation Coefficients

If we want to compute a downsampled correlation or autocorrelation coefficients, we can also use the polyphase representation. Take the autocorrelation coefficients $r_{xx}(n) = x(n) * x^*(-n)$ of a causal signal $x(n)$ (meaning it starts at some point in time, here at $n = 0$, such that the z-transform sums can have infinity on both sides). In (1.12) we saw that this corresponds to the power spectrum

$$R_{xx}(z) = X(z) \cdot X^*(z^{-1})$$

We can view $x(n)$ as the signal and $x^*(-n)$ as the filter. If we want to obtain the downsampled autocorrelation coefficients $r_{xx}^{\downarrow N}(m)$, this is equivalent to filter $x(n)$ with filter $x^*(-n)$ and then downsample the output. We can implement it more efficiently using the polyphase representation. We again take the polyphase components of $x(n)$ in the z-domain, with N phases, as

$$X_n(z) = \sum_{m=-\infty}^{\infty} x(mN + n)z^{-m}$$

and our filters polyphase components are

$$H_n(z) = \sum_{m=-\infty}^{\infty} x^*(-mN - n)z^{-m} = X_{-n}^*(z^{-1})$$

Using the polyphase representation, we can now write this filtering and downsampling by N operation in the z-domain as multiplication of the two polyphase vectors, as in (1.29),

$$R_{xx}^{\downarrow N}(z) = \sum_{n=0}^{N-1} X_n(z) \cdot H_n(z) = \sum_{n=0}^{N-1} X_n(z) \cdot X_{-n}^*(z^{-1})$$

Also observe that $X_{-n}^*(z^{-1}) = X_{N-1-n}^*(z^{-1})$. An example for the case $N = 2$ is

$$R_{xx}^{\downarrow 2}(z) = X_0(z) \cdot X_1^*(z^{-1}) + X_1(z) \cdot X_0^*(z^{-1}) \tag{1.41}$$

1.11.4 Parseval's Theorem for the Polyphase Representation

Parseval's theorem states that the signal power we compute in the time domain as its energy $\sum_{n=0}^{\infty} |x(n)|^2$) is identical to the power we compute in its frequency domain,

$$\sum_{n=0}^{\infty} |x(n)|^2 = \frac{1}{2\pi} \int_{\omega=0}^{2\pi} |X(e^{j\omega})|^2 d\omega$$

In Eq. (1.28) we see that we can rewrite our signal in terms of its polyphase components. In DTFT form and with our polyphase representation the spectral components are

$$X(e^{j\omega}) = \sum_{n=0}^{N-1} e^{-j\omega n} \cdot X_n^{\downarrow N}(e^{j\omega N})$$

Hence we get

$$\sum_{n=0}^{\infty} |x(n)|^2 = \frac{1}{2\pi} \int_{\omega=0}^{2\pi} \left| \sum_{n=0}^{N-1} e^{-j\omega n} \cdot X_n^{\downarrow N}(e^{j\omega N}) \right|^2 d\omega \tag{1.42}$$

1.12 Perfect Reconstruction

Going back to filter banks, our goal is now to find the filter bank impulse responses $h_k(n)$ and $g_k(n)$ such that the output of the synthesis filter bank is identical to the input of the analysis filter bank, except for some *system delay* n_d,

$$\hat{x}(n) = x(n - n_d)$$

The system delay results from the filtering operations in the analysis and synthesis filter bank from using causal filters, as we will see later.

To obtain the conditions for perfect reconstruction, we take a look at the output of the synthesis filter bank, using the representation using polyphase matrices,

$$\hat{\mathbf{X}}(z) = \mathbf{Y}(z) \cdot \mathbf{G}(z) = \mathbf{X}(z) \cdot \mathbf{H}(z) \cdot \mathbf{G}(z) \tag{1.43}$$

Perfect reconstruction means that the output of the synthesis filter bank $(\hat{\mathbf{X}}^T(z))$ is identical to the input of the analysis filter bank $(\mathbf{X}^T(z))$ except for a delay. For simplicity we restrict the delay introduced by the filters to be a multiple d of the blocksize N. Hence perfect reconstruction results if

$$\mathbf{G}(z) = z^{-d} \mathbf{H}^{-1}(z) \tag{1.44}$$

Here we can see that the design of the synthesis filter bank for perfect reconstruction "only" requires a matrix inversion of the analysis polyphase matrix $\mathbf{H}(z)$. The remaining problem is: how can we invert a matrix of polynomials, and preferably in a way that we obtain FIR filters for the synthesis if we have FIR filters in the analysis filter bank?

A simple approach for block transform matrices was to make them orthogonal. This leads to analysis and synthesis impulse responses, which are identical, except that the analysis impulse response is the time reversed of the synthesis impulse response. The analog for the polyphase description are the so-called **para-unitary** polyphase matrices [4], which have the property that their inverse is the transpose, with z replaced by z^{-1}, or

$$\mathbf{H}^{-1}(z) = \mathbf{H}^T(z^{-1}). \tag{1.45}$$

Hence, if we have analysis filters which lead to para-unitary polyphase matrices, the syntheses polyphase matrix can simply be obtained with the above equation. It also leads the analysis and synthesis filter impulse responses which are **time-reversed** versions of each other.

With the block transforms we also already saw that a very simple approach to obtain an inverse is to simply avoid polynomial entries in the polyphase matrix. We achieve this by restricting the length of our impulse responses to be maximally as long as one block (which is equal to the downsampling rate and the number of subbands N), hence $L \leq N$. In this way, the polyphase components $H_{k,i}(z)$ in (1.34) are simply scalars, since the sum has only one entry at $m = 0$, and the polyphase matrix turns into a transform matrix.

But the purpose of the polyphase representation was to be able to obtain longer filters than with block transforms, to obtain overlapping filter impulse responses without losing critical sampling. Hence we take a look at the next case, where we extend the filter length to $L \leq 2N$, and we obtain a so-called 50% overlap between neighbouring filter impulse responses. A widely used example is the so-called Modified Discrete Cosine Transform, or MDCT.

1.12.1 MDCT

The MDCT is an example for a so-called modulated filter bank, where the subband filters are obtained by modulating a baseband lowpass filter [9, 12, 16] (see also "Lapped Orthogonal Transforms", LOT [17]). It is used in the MPEG1/2 Layer 3 and MPEG2/4 AAC standards [3]. Modulation means the multiplication of an impulse response with a function (usually a periodic function), here a cosine function. In this way we can obtain the different bandpass filters from a baseband prototype. In the context of the MDCT, often the concept of a so-called window function is used. It is the time-reversed baseband prototype function $h(n)$ or $g(n)$. The concept of a window function is used when the MDCT is viewed as a transform. In this case the time index of the window function has the same direction (or sign) as the signal time index. In the section about the connection between transforms and filter banks (1.10) we saw that the equivalent impulse responses of an analysis transform, viewed as a filter bank, are the time reversed columns of the transform matrix. Hence window functions and baseband prototype functions are time reversed versions of each other. If we start with a baseband prototype filter $h(n)$, the N impulse responses $h_k(n)$ of an MDCT analysis filter bank are defined as

$$h_k(n) = -h(n) \cdot \sqrt{\frac{2}{N}} \cos\left(\frac{\pi}{N}\left(k + \frac{1}{2}\right)\left(n \pm \frac{N}{2} + \frac{1}{2}\right)\right) \qquad (1.46)$$

for the subbands $k = 0, \ldots, N-1$ and the time index $n = 0, \ldots, 2N-1$, meaning we have filters of length $LN = 2N$. This means we now have an impulse response of length of 2 blocks of N samples.

The synthesis filters are produced from the baseband prototype $g(n)$,

$$g_k(n) = g(n) \cdot \sqrt{\frac{2}{N}} \cos\left(\frac{\pi}{N}\left(k + \frac{1}{2}\right)\left(n \mp \frac{N}{2} + \frac{1}{2}\right)\right)$$

This filter bank is known to achieve perfect reconstruction, if the baseband prototype filters $h(n)$ and $g(n)$ are chosen right, despite the critical sampling and the longer filters. Now how do we choose the prototype filters? To find an answer, we construct the corresponding polyphase matrices. For the analysis we obtain the elements of its polyphase matrix from Eq. (1.34) as

$$H_{n,k}^{\downarrow N}(z) = \sum_{m=0}^{\infty} h_k(mN + n)z^{-m}.$$

and for the synthesis polyphase matrix we get the elements of the synthesis polyphase matrix with Eq. (1.37)

$$G_{k,n}^{\downarrow N}(z) = \sum_{m=0}^{\infty} g_k(mN + n)z^{-m}.$$

Here we can see that these elements of the polyphase matrices are polynomials of only first order, since the sums only run from 0 to 1.

In this way, the analysis polyphase matrix for the MDCT becomes

$$\mathbf{H}_{MDCT}(z) =$$

$$= \begin{bmatrix} h_0(N-1) + h_0(2N-1)z^{-1} & \cdots & h_{N-1}(N-1) + h_{N-1}(2N-1)z^{-1} \\ h_0(N-2) + h_0(2N-2)z^{-1} & & \vdots \\ \vdots & \ddots & \\ & \vdots & \vdots & \\ h_0(0) + h_0(N)z^{-1} & \cdots & h_{N-1}(0) + h_{N-1}(N)z^{-1} \end{bmatrix} \quad (1.47)$$

Observe that the second block of our impulse response appears as coefficients of z^{-1}. If we have this polyphase matrix given, then we can extract h_0 from its first column by flipping the column up/down and then appending the column with the coefficients of z^{-1}. We can then compute the baseband prototype filter $h(n)$ by dividing h_0 by its modulation function (1.46),

$$h(n) = h_0(n) / \left(\sqrt{\frac{2}{N}} \cos\left(\frac{\pi}{N}\left(\frac{1}{2}\right)\left(n \pm \frac{N}{2} + \frac{1}{2}\right)\right)\right)$$

The synthesis polyphase matrix for the MDCT becomes

$$\mathbf{G}_{MDCT}(z) =$$

$$= \begin{bmatrix} g_0(0) + g_0(N)z^{-1} & \cdots & g_0(N-1) + g_0(2N-1)z^{-1} \\ g_1(0) + g_1(N)z^{-1} & & \vdots \\ \vdots & \ddots & \\ & \vdots & \vdots & \\ g_{N-1}(0) + g_{N-1}(N)z^{-1} & \cdots & g_{N-1}(N-1) + g_{N-1}(2N-1)z^{-1} \end{bmatrix}$$

We can now use the periodicities and symmetries of the cosine modulation function to simplify these matrices. Observe that the modulation function has similarity to the DCT type 4, as shown in Eq. (1.22). We get an interesting effect if we multiply the MDCT analysis polyphase matrix from the right with the inverse of the DCT4 transform matrix $\mathbf{T}$. The result is a sparse matrix with a particular form, which we call a "folding matrix", $\mathbf{F_a}(z)$. In the first case of $+N/2$ in the modulation function it is

$$\mathbf{F_a}(z) := \mathbf{H}_{MDCT}(z) \cdot \mathbf{T}^{-1} =$$

$$= \begin{bmatrix} 0 & & -z^{-1}h(2N-1) & -h(N-1) & & 0 \\ & \ddots & & & \ddots & \\ -z^{-1}h(1.5N) & & 0 & & & -h(0.5N) \\ -z^{-1}h(1.5N-1) & & & 0 & & h(0.5N-1) \\ & \ddots & & & \ddots & \\ 0 & & -z^{-1}h(N) & h(0) & & 0 \end{bmatrix}$$

$$(1.48)$$

In the second case, with $-N/2$ in the modulation function, it is

$$\mathbf{F_a}(z) := \mathbf{H}_{MDCT}(z) \cdot \mathbf{T}^{-1} =$$

$$= \begin{bmatrix} 0 & & h(N-1) & -z^{-1}h(2N-1) & & 0 \\ & \ddots & & & \ddots & \\ h(0.5N) & & 0 & & & -z^{-1}h(1.5N) \\ h(0.5N-1) & & & 0 & & z^{-1}h(1.5N-1) \\ & \ddots & & & \ddots & \\ 0 & & h(0) & z^{-1}h(N) & & 0 \end{bmatrix} \quad (1.49)$$

Observe that in this case the delays appear on the other side.

Here we can see that we now obtained a sparse matrix with only $2N$ non-zero entries! This means we obtain the MDCT polyphase matrix by multiplying this sparse folding matrix $\mathbf{F_a}(z)$ with the DCT4 transform matrix. We can see that each element or sample of the baseband prototype $h(n)$ is multiplied with specific entries of the DCT4 transform matrix $\mathbf{T}$. What is the connection of the MDCT modulation function to the DCT4? When we look at the MDCT modulation function,

$$\cos\left(\frac{\pi}{N}\left(k+\frac{1}{2}\right)\left(n \pm \frac{N}{2} + \frac{1}{2}\right)\right)$$

and compare it with the DCT4 transform equivalent modulation function,

$$\cos\left(\frac{\pi}{N}\left(k+\frac{1}{2}\right)\left(n+\frac{1}{2}\right)\right)$$

we can see that the MDCT modulation function is basically a time shifted version of the DCT4 modulation function, and it extends over $2N$ samples instead of just N samples. To see what is happening with the modulation function outside the $n = 0, \ldots, N-1$ range of the DCT4 modulation function, one can use the periodicities and symmetries of the cosine function to map the outside range back into the original range. This means that all occurring values of the MDCT modulation function occur also in the DCT4 transform matrix, and hence can be mapped to them by using a suitable position in our sparse folding matrix $\mathbf{F_a}(z)$. Negative time indices n for the DCT4 modulation can occur since we have a time shift of $-N/2$ for the MDCT modulation function. We can map them back to positive n values with

$$\cos\left(\frac{\pi}{N}\left(k+\frac{1}{2}\right)\left(-n+\frac{1}{2}\right)\right) = \cos\left(\frac{\pi}{N}\left(k+\frac{1}{2}\right)\left(n-\frac{1}{2}\right)\right) =$$

$$= \cos\left(\frac{\pi}{N}\left(k + \frac{1}{2}\right)\left(n - 1 + \frac{1}{2}\right)\right)$$

meaning instead of reducing the index n below zero we just increase it accordingly. We can also obtain values of n which are larger than $N - 1$ since for the MDCT the time index range is $n = 0, \ldots, 2N - 1$. We can map those values back into the range below N by using

$$\cos\left(\frac{\pi}{N}\left(k + \frac{1}{2}\right)\left(N + n + \frac{1}{2}\right)\right) = -\cos\left(\frac{\pi}{N}\left(k + \frac{1}{2}\right)\left(N - 1 - n + \frac{1}{2}\right)\right)$$

This means instead of increasing the index n above N we just decrease it accordingly and flip the sign of the cosine function.

This shows why the matrix $\mathbf{F_a}(z)$ results in its specific shape.

Another advantage is that each entry of the folding matrix $\mathbf{F_a}(z)$ has only one variable, a sample of the baseband prototype impulse response. Further we can see that only the left half of the matrix contains a multiplication of z^{-1}, a simple multiplication such that we can multiply it from a separate matrix with the delays from the right. Let's define a "delay matrix" as

$$\mathbf{D}(z) = \begin{bmatrix} z^{-1} & & & & & \\ & \ddots & & & 0 & \\ & & z^{-1} & & & \\ & & & 1 & & \\ & 0 & & & \ddots & \\ & & & & & 1 \end{bmatrix} \tag{1.50}$$

Hence we can write the matrix $\mathbf{F_a}(z)$ of (1.53) as a product of this delay matrix with a folding matrix without dependency on z^{-1}, which we call $\mathbf{F_a}$ (the missing dependency on z here now signifies that we have multiplied out the delays) and which only contains the samples of the baseband prototype function,

$$\mathbf{F_a} = \begin{bmatrix} 0 & & h(2N-1) & h(N-1) & & 0 \\ & \cdot^{\cdot^{\cdot}} & & & \ddots & \\ h(1.5N) & & 0 & & & h(0.5N) \\ h(1.5N-1) & & & 0 & & -h(0.5N-1) \\ & \cdot^{\cdot^{\cdot}} & & & \cdot^{\cdot^{\cdot}} & \\ 0 & & h(N) & -h(0) & & 0 \end{bmatrix} \tag{1.51}$$

Then we can write the MDCT analysis polyphase matrix $\mathbf{H}_{MDCT}(z)$ as the product of the sparse matrices (which are efficient to implement) $\mathbf{F_a}$ and $\mathbf{D}(z)$, and the transform matrix $\mathbf{T}$, which can also be efficiently implemented, using a fast DCT or also FFTs [18],

$$\mathbf{H}_{MDCT}(z) = \mathbf{F_a} \cdot \mathbf{D}(z) \cdot \mathbf{T} \tag{1.52}$$

For the synthesis side we can use the same approach. For the synthesis we will use the inverse transform, hence we use the inverse of the inverse, which is the transform itself, to obtain a sparse matrix,

$$\mathbf{T} \cdot \mathbf{G}_{MDCT}(z) =$$

$$
= \begin{bmatrix}
0 & & g(0.5N-1) & g(0.5N) & & & 0 \\
& \cdot\cdot\cdot & & & & \cdot\cdot & \\
g(0) & & 0 & & & & g(N-1) \\
z^{-1}g(N) & & & 0 & & & -z^{-1}g(2N-1) \\
& \cdot\cdot & & & & \cdot\cdot\cdot & \\
0 & & z^{-1}g(1.5N-1) & -z^{-1}g(1.5N) & & & 0
\end{bmatrix} \quad (1.53)
$$

Using this matrix we define a folding matrix $\mathbf{F_s}$ as

$$
\mathbf{F_s} = \begin{bmatrix}
0 & & g(0.5N-1) & g(0.5N) & & & 0 \\
& \cdot\cdot\cdot & & & & \cdot\cdot & \\
g(0) & & 0 & & & & g(N-1) \\
g(N) & & & 0 & & & -g(2N-1) \\
& \cdot\cdot & & & & \cdot\cdot\cdot & \\
0 & & g(1.5N-1) & -g(1.5N) & & & 0
\end{bmatrix}
$$

The delay elements are now on the lower half of the matrix, which means we can obtain them by multiplying the folding matrix $\mathbf{F_s}$ from the left with $z^{-1} \cdot \mathbf{D}^{-1}(z)$. We can now also write the synthesis polyphase matrix as a product of two sparse matrices and a fast implementable transform matrix,

$$\mathbf{G}_{MDCT}(z) = \mathbf{T}^{-1} \cdot z^{-1} \cdot \mathbf{D}^{-1}(z) \cdot \mathbf{F_s} \quad (1.54)$$

Perfect Reconstruction of the MDCT We can now use Eq. (1.52) and (1.54) to compute the direct concatenation of the analysis and synthesis filter bank as a product of their polyphase matrices. For perfect reconstruction, this should only correspond to a delay. We obtain

$$\mathbf{H}_{MDCT}(z) \cdot \mathbf{G}_{MDCT}(z) = \mathbf{F_a} \cdot \mathbf{D}(z) \cdot \mathbf{T} \cdot \mathbf{T}^{-1} \cdot z^{-1} \cdot \mathbf{D}^{-1} \cdot \mathbf{F_s} =$$

$$= \mathbf{F_a} \cdot z^{-1} \cdot \mathbf{F_s}$$

Here we can see that we indeed obtain perfect reconstruction, if we choose $\mathbf{F_s} = \mathbf{F_a}^{-1}$. In that case all that is left is the delay z^{-1}, which is here at the lower sampling rate, meaning it corresponds to a delay of 1 block of N samples.

Since $\mathbf{F_a}$ only consists of scalars, its inverse is easily obtained, for instance numerically,

1 Filter Banks

or algebraically by seeing that it consists of a set of nested 2×2 matrices,

$$
\mathbf{F_a} = \begin{bmatrix} 0 & & & \cdots & \ddots & & & 0 \\ & h(2N-1-n) & & & & h(N-1-n) & & \\ \ddots & & & 0 & & & & \ddots \\ \ddots & & & & 0 & & & \ddots \\ & h(N+n) & & & & -h(n) & & \\ 0 & & & \ddots & \ddots & & & 0 \end{bmatrix}
$$

with $n = 0, \ldots, N/2 - 1$. Hence these 2×2 submatrices are of the form

$$
\begin{bmatrix} h(2N-1-n) & h(N-1-n) \\ h(N+n) & -h(n) \end{bmatrix} \tag{1.55}
$$

For the synthesis the corresponding 2x2 matrix is

$$
\begin{bmatrix} g(n) & g(N-1-n) \\ g(N+n) & -g(2N-1-n) \end{bmatrix}
$$

The inverse of the analysis submatrices is [19]

$$
\frac{1}{h(2N-1-n)h(n) + h(N-1-n)h(N+n)} \begin{bmatrix} h(n) & h(N-1-n) \\ h(N+n) & -h(2N-1-n) \end{bmatrix}
$$

Observe that the denominator is the negative determinant of the submatrix. Hence by setting $\mathbf{F_s} = \mathbf{F_a}^{-1}$ we obtain

$$
\begin{bmatrix} g(n) & g(N-1-n) \\ g(N+n) & -g(2N-1-n) \end{bmatrix} =
$$

$$
= \frac{1}{h(2N-1-n)h(n) + h(N-1-n)h(N+n)} \begin{bmatrix} h(n) & h(N-1-n) \\ h(N+n) & -h(2N-1-n) \end{bmatrix}
$$

This means that for every window function $h(n)$ for which the determinant of the submatrices is unequal to zero, we can obtain a synthesis window function $g(n)$ for perfect reconstruction! A particularly easy choice is to design $h(n)$ such that this nominator, the **negative determinant becomes** ± 1,

$$
h(2N-1-n)h(n) + h(N-1-n)h(N+n) = \pm 1.
$$

In that case we get

$$
g(n) = \pm h(n),
$$

and we obtain a "symmetry" in the sense that the **analysis and synthesis prototype filters are identical, except for the sign**. This is practical because we only need to design one side of our filter bank, for instance the analysis filter bank prototype. The other side is then just the same, and we know that it also has the same filter characteristics. We can obtain this, for instance, by specifying in the above matrix $h(n), h(N+n), h(N-1-n)$, and compute $h(2N-1-n)$ from this requirement. We have

$$\pm 1 = h(2N-1-n)h(n) + h(N-1-n)h(N+n)$$

$$\frac{\pm 1 - h(N-1-n)h(N+n)}{h(n)} = h(2N-1-n)$$

for $n = 0, \ldots, N/2$. Observe that this does not necessarily lead to para-unitary filter banks. From Eq. (1.45) we see that we obtain a para-unitary filter bank if we set

$$\mathbf{F_a}^{-1} = \mathbf{F_a}^T$$

(and replace z by z^{-1} in $\mathbf{D}(z)$ which was the inverse anyway, and leave the DCT transform matrix $\mathbf{T}$ unchanged since it is symmetric). This results in

$$\frac{1}{h(2N-1-n)h(n) + h(N-1-n)h(N+n)} \begin{bmatrix} h(n) & h(N-1-n) \\ h(N+n) & -h(2N-1-n) \end{bmatrix} =$$

$$= \begin{bmatrix} h(2N-1-n) & h(N+n) \\ h(N-1-n) & -h(n) \end{bmatrix}$$

for $n = 0, \ldots, N/2 - 1$. Here we can see that we obtain a para-unitary filter bank if we again set the negative determinant to one, and if we set $h(n) = h(2N-1-n)$ and $h(N+n) = h(N-1-n)$ for $n = 0, \ldots, N/2 - 1$, or simply $h(n) = h(2N-1-n)$ for $n = 0, \ldots, N-1$. This means for a **para-unitary polyphase matrix**, our baseband prototype impulse response $h(n)$ needs to be **symmetric around its centre**, meaning it needs to be identical to its time reversed version to obtain a para-unitary filter bank! Since we also need the property of the negative determinant of the submatrices to be one, para-unitarity also means that analysis and synthesis prototype filters are identical.

A simple and widespread example for this choice is the so-called sine-window. It is defined as

$$h(n) = \sin\left(\frac{\pi}{2N}(n+0.5)\right)$$

for $n = 0, \ldots, 2N-1$, which leads to a para-unitary filter bank and polyphase matrix, because the window function is symmetric around its centre and it also leads to negative determinants of the submatrices which are one, hence analysis and synthesis share this same prototype filter.

1.12.2 Python Example

In Python we can use 3-dimensional matrices to represent polyphase matrices. The m'th position in the third dimension then stores the coefficient of z^{m-1}. We then need a function to multiply two polyphase matrices. We can write one by implementing a convolution of matrices, by viewing a polyphase matrix as a polynomial of matrices. An example is the following function,

```
import numpy as np
def polmatmult( A,B ):
    """polmatmult(A,B)
    multiplies two polynomial matrices (arrays) A and B, where each matrix entry is a polynomial.
    Those polynomial entries are in the 3rd dimension
    The thirs dimension can also be interpreted as containing the (2D) coefficient
    exponent of z^-1.
    Result is C=A*B;"""
    [NAx, NAy, NAz] = np.shape(A);
    [NBx, NBy, NBz] = np.shape(B);
    #Degree +1 of resulting polynomial, with NAz-1 and NBz-1 being the degree of the...
    Deg = NAz + NBz -1;
    C = np.zeros((NAx,NBy,Deg));
    #Convolution of matrices:
    for n in range(0,(Deg)):
        for m in range(0,n+1):
            if ((n-m)<NAz and m<NBz):
                C[:,:,n] = C[:,:,n]+ np.dot(A[:,:,(n-m)],B[:,:,m]);
    return C
```

We store it as file with name "polmatmult.py" (we will do the same with the following functions). The matrix $\mathbf{D}(z)$ can then be generated by the function

```
import numpy as np
def Dmatrix(N):
    """produces a delay matrix D(z), which has delay z^-1 on the upper half of its diagonal
    in a 3D polynomial representation (exponents of z^-1 are in the third dimension)
    N is number of subbands and size of the polynomial matrix (NxN)
    N is even"""
    D=np.zeros((N,N,2));
    D[:,:,0] = np.diag(np.append(np.zeros((1,int(N/2))), np.ones((1,int(N/2)))));
    D[:,:,1] = np.diag(np.append(np.ones((1,int(N/2))), np.zeros((1,int(N/2)))));
    return D;
```

For instance, for $N = 4$ we obtain this matrix as

```
D=Dmatrix(4)
D[:,:,0]

array([[ 0.,   0.,   0.,   0.],
       [ 0.,   0.,   0.,   0.],
       [ 0.,   0.,   1.,   0.],
       [ 0.,   0.,   0.,   1.]])
```

```
D[:,:,1]
```

```
array([[ 1.,   0.,   0.,   0.],
       [ 0.,   1.,   0.,   0.],
       [ 0.,   0.,   0.,   0.],
       [ 0.,   0.,   0.,   0.]])
```

We would like to have the symmetry between analysis and synthesis filter bank such that they have the same baseband impulse responses. We saw that we can obtain this by requiring the submatrices of $\mathbf{F}$ to be -1, hence its last $N/2$ coefficients can be computed from its first 1.5N coefficients, included in the function generation the matrix $\mathbf{F}$,

```
from __future__ import print_function
import numpy as np
def symFmatrix(f):
    """produces a diamond shaped folding matrix F from the coefficients f
    (f is a 1−d array)
    which leads to identical analysis and synthesis baseband impulse responses
    Hence has det 1 or −1
    If N is number of subbands, then f is a vector of size 1.5*N coefficients.
    N is even
    returns: F of shape (N,N,1)
    """
    sym=1.0; #The kind of symmetry: +−1
    N = int(len(f)/1.5);
    F=np.zeros((N,N,1))
    F[0:int(N/2),0:int(N/2),0]=np.fliplr(np.diag(f[0:int(N/2)]))
    F[int(N/2):N,0:int(N/2),0]=np.diag(f[int(N/2):N])
    F[0:int(N/2),int(N/2):N,0]=np.diag(f[N:(N+int(N/2))])
    ff = np.flipud((sym*np.ones((int(N/2))) − (f[N:(int(1.5*N))])*f[N−1:int(N/2)−1:−1])/
          f[0:int(N/2)])
    F[int(N/2):N,int(N/2):N,0]=−np.fliplr(np.diag(ff))
```

For the MDCT with the Sine window, the coefficients f are the coefficients of this Sine window. Now assume we would like to have a 4-band MDCT filter bank, hence $N = 4$. We only need to specify the first 1.5N elements. The diamond shaped folding matrix for the analysis $\mathbf{F_a}$ is then obtained with the following ipython code:

```
ipython --pylab
from symFmatrix import *

N=4;
n= arange(0,1.5*N)
f=sin(pi/(2*N)*(n+0.5))
Fa=symFmatrix(f)
```

```
Fa[:,:,0]
Out:
array([[ 0.        ,  0.19509032,  0.98078528,  0.        ],
       [ 0.55557023,  0.        ,  0.        ,  0.83146961],
       [ 0.83146961,  0.        , -0.        , -0.55557023],
       [ 0.        ,  0.98078528, -0.19509032, -0.        ]])
```

The multiplication of this folding matrix $\mathbf{F_a}$ with the delay matrix $\mathbf{D}(z)$ is now

```
from Dmatrix import *
from polmatmult import *
D=Dmatrix(N)
Faz=polmatmult(Fa,D)
Faz[:,:,0]
Out:
array([[ 0.        ,  0.        ,  0.98078528,  0.        ],
       [ 0.        ,  0.        ,  0.        ,  0.83146961],
       [ 0.        ,  0.        ,  0.        , -0.55557023],
       [ 0.        ,  0.        , -0.19509032,  0.        ]])
Faz[:,:,1]
Out:
array([[ 0.        ,  0.19509032,  0.        ,  0.        ],
       [ 0.55557023,  0.        ,  0.        ,  0.        ],
       [ 0.83146961,  0.        ,  0.        ,  0.        ],
       [ 0.        ,  0.98078528,  0.        ,  0.        ]])
```

Optimization Example

In the previous example we took the Sine window for our coefficients. Instead of given coefficients, we can obtain them using numerical optimization. For that we need an error function which we can minimize. First we need a function which computes the baseband prototype function given a set of coefficients, using Eq. (1.47), by extracting the impulse response of the lowest subband ($k = 0$) and dividing it by the modulation function. We use the following function and store it in the file "Fa2h.py".

def Fa2h(Fa):
    """ Function extracts analysis baseband impulse response (reverse window function)
    from the folding matrix Fa of a cosine modulated filter bank.
    """
    import numpy as np
    from polmatmult **import** polmatmult
    [N,y,blocks] = np.shape(Fa)
    h0 = np.zeros(blocks * N)
    #First column of DCT−4:
    T = np.zeros((N,1,1))

```
T[:,0,0]=np.cos(np.pi/N*(0.5)*(np.arange(N)+0.5))
#Compute first column of Polyphase matrix Pa(z):
Pa = polmatmult(Fa,T)
#Extract impulse response h0(n):
for m in range(blocks):
    h0[m*N+np.arange(N)] = np.flipud(Pa[:,0,m])
#Baseband prototype h(n), divide by modulation func.:
#h = h0 / np.cos(np.pi/N*0.5*(np.arange(blocks*N)+ 0.5+(N/2)))
h = −h0 / np.cos(np.pi/N*0.5*(np.arange(blocks*N−1,−1,−1)+ 0.5−(N/2)))
return h;
```

Note that the modulation function we divide by always starts with the same phase from its end, which becomes important when we extend the filter length later. This also corresponds to the time direction of viewing the prototype as a window function.

Next we write the following error function, which produces a measure about how close we are to a given ideal specification in the frequency domain, and store it in the file "optimfuncMDCT.py".

```
def optimfuncMDCT(x, N):
    """Computes the error function for the filter bank optimization
    for coefficients x, a 1−d array, N: Number of subbands"""
    import numpy as np
    import scipy.signal as sig
    from polmatmult import polmatmult
    from Dmatrix import Dmatrix
    from symFmatrix import symFmatrix
    from Fa2h import Fa2h

    #x = np.transpose(x)
    Fa = symFmatrix(x)
    D = Dmatrix(N)
    Faz = polmatmult(Fa,D)
    h = Fa2h(Faz)
    h = np.hstack(h)
    w, H = sig.freqz(h,1,1024)
    pb = int(1024/N/2)
    Hdes = np.concatenate((np.ones((pb,1)) , np.zeros(((1024−pb, 1)))), axis = 0)
    tb = np.round(pb)
    weights = np.concatenate((np.ones((pb,1)) , np.zeros((tb, 1)),
    1000*np.ones((1024−pb−tb,1))), axis = 0)
    err = np.sum(np.abs(H−Hdes)*weights)
    return err
```

Now we can start the actual optimization. We use the Python library "scipy.optimize" and its function "minimize" for it (we need scipy version 0.19.1 or later for it to work properly, the version is shown in Python with `scipy.__version__`). We add the optimization in the "main" routine of the file "optimfuncMDCT.py". After the optimization the found coefficients are stored in the text file "MDCTcoeff.txt", and the resulting baseband impulse response and the magnitude of its frequency response are computed.

```python
if __name__ == '__main__': #run the optimization
    import numpy as np
    import scipy as sp
    import scipy.optimize
    import scipy.signal
    import matplotlib.pyplot as plt
    from symFmatrix import symFmatrix
    from polmatmult import polmatmult
    from Dmatrix import Dmatrix
    from Fa2h import Fa2h

    N=4
    #Start optimization with some starting point:
    x0 = -np.random.rand(int(1.5*N))
    print("starting_error=", optimfuncMDCT(x0, N)) #test optim. function
    xmin = sp.optimize.minimize(optimfuncMDCT, x0, args=(N,), options={'disp':True})
    print("optimized_coefficients=", xmin.x)
    np.savetxt("MDCTcoeff.txt", xmin.x)
    print("error_after_optim.=", optimfuncMDCT(xmin.x, N))
    #Baseband Impulse Response:
    Fa = symFmatrix(xmin.x)
    Faz = polmatmult(Fa, Dmatrix(N))
    h = Fa2h(Faz)
    print("h=", h)
    plt.plot(h)
    plt.xlabel('Sample')
    plt.ylabel('Value')
    plt.title('Baseband_Impulse_Response_of_our_Optimized_MDCT_Filter_Bank')
    plt.figure()
    #Magnitude Response:
    w,H=sp.signal.freqz(h)
    plt.plot(w,20*np.log10(abs(H)))
    plt.axis([0, 3.14, -60,20])
    plt.xlabel('Normalized_Frequency')
    plt.ylabel('Magnitude_(dB)')
    plt.title('Mag._Frequency_Response_of_the_MDCT_Filter_Bank')
    plt.show()
```

Then we run it with

```
python optimfuncMDCT.py
```

Observe that the obtained baseband prototype impulse response, in Fig. 1.19, is similar to the sine window, but not exactly the same. The resulting magnitude of the frequency response can be seen in Fig. 1.20. We can see that the 3 dB bandwidth of the passband is about $\pi/8$, as desired, and that the first sidelobe at about normalized frequency 1.4 has an attenuation of about -25 dB relative to the pass band.

MDCT Filter Bank Implementation

To process an audio signal in the polyphase representation, we first need to turn it into a polyphase vector, using our 3-d tensor notation. This can be done with the following function:

```
from __future__ import print_function
import numpy as np

def x2polyphase(x,N):
    """Converts input signal x (a 1D array) into a polyphase row vector
    xp for blocks of length N, with shape: (1,N,#of blocks)"""
    import numpy as np
    #Convert stream x into a 2d array where each row is a block:
    #xp.shape : (y,x, #num blocks):
    x=x[:int(len(x)/N)*N] #limit signal to integer multiples of N
    xp=np.reshape(x,(N,-1),order='F') #order=F: first index changes fastest
    #add 0'th dimension for function polmatmult:
    xp=np.expand_dims(xp,axis=0)
    return xp
```

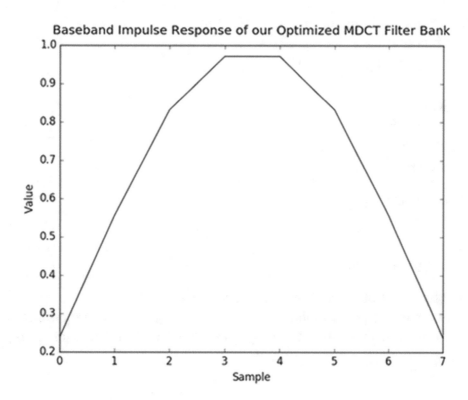

Figure 1.19: The baseband prototype impulse response resulting from our optimization.

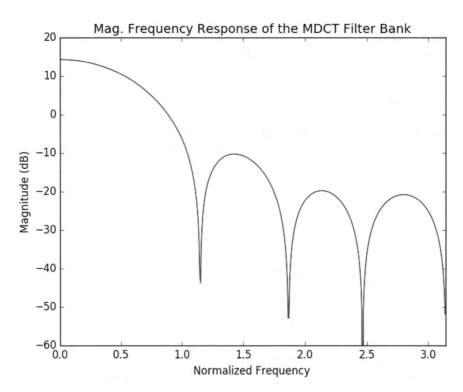

Figure 1.20: The baseband prototype frequency response resulting from our optimization.

For testing we add a main routine after the function. As input signal, we assume a simple ramp function.

```
if __name__ == '__main__':
    #testing:
    x=np.arange(0,9)
    xp=x2polyphase(x,4)
    print(xp[:,:,0], xp[:,:,1])
```

We let it run with

```
python x2polyphase.py
```

and obtain

```
[[0 1 2 3]] [[4 5 6 7]]
```

We see the 2 blocks of our signal, which is indeed what we expect.

We also need the reverse, a function which takes the polyphase representation of the signal and returns it to a stream, a 1-d array. We do that with the function "polyphase2x",

```
from __future__ import print_function
import numpy as np

def polyphase2x(xp):
```

"""Converts polyphase input signal xp (a row vector) into a contiguos row vector
For block length N, for 3D polyphase representation (exponents of z in the third
matrix/tensor dimension)"""

x=np.reshape(xp,(1,1,−1),order='F') *#order=F: first index changes fastest*
x=x[0,0,:]
return x

We also add a main routine for testing, using our ramp function,

if __name__ == '__main__':
 #testing:
 from x2polyphase **import** *
 x=np.arange(0,8)
 xp=x2polyphase(x,4)
 xrek=polyphase2x(xp)
 print(xrek)

and start it with

```
python polyphase2x.py
```

This yields

```
[0 1 2 3 4 5 6 7]
```

which is the reconstructed 1-d array of our ramp function, as we expected.

For the implementation of an MDCT filter bank it speeds up the running time if we use a fast implementation of the DCT4 (1.21), especially for large numbers of subbands N. Pythons scipy.fftpack already contains a fast DCT3, but not a fast DCT4. Fortunately we can implement a fast DCT4 using a fast DCT3.

The DCT3 has the following definition:

$$y_k(m) = \sqrt{\frac{1}{N}} \cdot x(mN) + \sqrt{\frac{2}{N}} \cdot \sum_{n=0}^{N-1} x(mN+n) \cdot \cos(\frac{\pi}{N}(k+0.5)(n)) \qquad (1.56)$$

Observe that it only differs from the DCT4 in that it has a time index n without the addition of 0.5. We can now produce this missing addition by upsampling the block x to be transformed locally with phase 1, meaning,

$$xu(1 + 2 \cdot n) = x(mN + n)$$

(the in-between samples are zero), and use that as the input to the DCT3. Since we get a sequence of twice the length, N becomes $2N$, and we only need the lower half of the produced transform coefficients. Since the first sample of our sequence is zero, the first term of the DCT3 disappears

$$y_k(m) = \sqrt{\frac{2}{2N}} \cdot \sum_{n=0}^{2N-1} xu(n) \cdot \cos(\frac{\pi}{2N}(k+0.5)(n)) = \qquad (1.57)$$

$$= \sqrt{\frac{1}{2}} \sqrt{\frac{2}{N}} \cdot \sum_{n=0}^{N-1} xu(1+2n) \cdot \cos(\frac{\pi}{2N}(k+0.5)\frac{1}{2}(1+2n)) =$$

$$= \sqrt{\frac{1}{2}} \sqrt{\frac{2}{N}} \cdot \sum_{n=0}^{N-1} x(mN+n) \cdot \cos(\frac{\pi}{2N}(k+0.5)(n+0.5)) \qquad (1.58)$$

each time with $k = 0, \ldots, N-1$. The first line, Eq. (1.57), is the DCT3 of our upsampled signal xu, and in the last line, Eq. (1.58), we see that it is identical to $\sqrt{\frac{1}{2}}$ times the DCT4 of our signal x! Hence we can apply the DCT3 to our upsampled signal and multiply the result with $\sqrt{2}$ to obtain the DCT4 of the signal. The corresponding Python function is

```
import scipy.fftpack as spfft
#The DCT4 transform:
def DCT4(samples):
    #Argument: 3−D array of samples, shape (y,x,# of blocks), each row correspond to 1 row
    #to apply the DCT to.
    #Output: 3−D array where each row ist DCT4 transformed, orthonormal.
    import numpy as np
    #use a DCT3 to implement a DCT4:
    r,N,blocks=samples.shape
    samplesup=np.zeros((1,2*N,blocks))
    #upsample signal:
    samplesup[0,1::2,:]=samples
    y=spfft.dct(samplesup,type=3,axis=1,norm='ortho')*np.sqrt(2)
```

Since the DCT4 transform matrix is symmetric and orthonormal, the inverse DCT4 transform is identical to its forward transform, which simplifies our work since we only need this one function. We store this function in the file "DCT4.py". The following functions are stored in the file "MDCTfb.py".

The next function is for the complete MDCT analysis filter bank, which is built from our previous functions, and implements Eq. (1.52),

```
from DCT4 import *
from symFmatrix import symFmatrix
from Dmatrix import Dmatrix
from polmatmult import polmatmult
from x2polyphase import *
def MDCTanafb(x,N,fb):
    #MDCT analysis filter bank.
    #Arguments: x: input signal, e.g. audio signal, a 1−dim. array
    #N: number of subbands
    #fb: coefficients for the MDCT filter bank, for the F matrix, np.array with 1.5*N coefficients.
    #returns y, consisting of blocks of subband in in a 2−d array of shape (N,# of blocks)

    Fa=symFmatrix(fb)
    D=Dmatrix(N)
    y=x2polyphase(x,N)
```

```
y=polmatmult(y,Fa)
y=polmatmult(y,D)
y=DCT4(y)
#strip first dimension:
y=y[0,:,:]
return y
```

The MDCT synthesis filter bank needs the causal inverse of the delay matrix (1.50), $z^{-1} \cdot \mathbf{D}^{-1}(z)$, which is obtained using the function

```
def Dinvmatrix(N):
    """produces a causal inverse delay matrix D^{-1}(z), which has
    delays z^-1 on the lower half in 3D polynomial representation (exponents of z^-1
    are in third dimension)
    N is the number of subbands and size of the polynomial matrix (NxN)
    N is even"""
    import numpy as np
    D = np.zeros((N,N,2))
    D[:,:,0] = np.diag((np.append(np.ones((1,int(N/2))),np.zeros((1,int(N/2))))))
    D[:,:,1] = np.diag((np.append(np.zeros((1,int(N/2))),np.ones((1,int(N/2))))))
    return D
```

The function for the MDCT synthesis filter bank now implements Eq. (1.54),

```
from Dinvmatrix import Dinvmatrix
from polyphase2x import *
def MDCTsynfb(y,fb):
    #MDCT synthesis filter bank.
    #Arguments: y: 2-d array of blocks of subbands, of shape (N, # of blokcs)
    #returns xr, the reconstructed signal, a 1-d array.

    N=y.shape[0]
    Fa=symFmatrix(fb)
    #invert Fa matrix for synthesis after removing last dim:
    Fs=np.linalg.inv(Fa[:,:,0])
    #add again last dimension for function polmatmult:
    Fs=np.expand_dims(Fs, axis=-1)
    Dinv=Dinvmatrix(N)

    #add first dimension to y for polmatmult:
    y=np.expand_dims(y,axis=0)
    xp=DCT4(y)
    xp=polmatmult(xp,Dinv)
    xp=polmatmult(xp,Fs)
    xr=polyphase2x(xp)
```

Finally for testing we again add a main routine, which uses the simple ramp function as input signal for the analysis filter bank, because corruptions of the signal are easy to spot if perfect reconstruction is broken. It then shows the resulting subbands as a 2-d image, which should show energy mostly in the lowest subband for this signal, and then plots the reconstructed ramp signal (Figs. 1.21, 1.22, and 1.23). Further the test routine tests the impulse response of the synthesis filter bank, by using a unit pulse in subband 0 (and zeros everywhere else) for the input of the synthesis filter bank, and plots its output, which then contains only the impulse response of subband 0 (Fig. 1.24). This is to check if the filter operations are working correctly. The impulse response should be a smooth function without discontinuities. If both, perfect reconstruction and filtering work correctly, this is a good indication that the filter bank implementation is correct. The program loads our optimized coefficients file "MDCTcoeff.txt" for a 4-band MDCT, but alternatively a sine window for an arbitrary number of subbands can be used, by uncommenting the line above (and commenting out the "loadtxt" line).

```python
#Testing:
if __name__ == '__main__':
    import numpy as np
    import matplotlib.pyplot as plt

    #Number of subbands:
    N=4
    D=Dmatrix(N)
    Dinv=Dinvmatrix(N)
    #Filter bank coefficients for sine window:
    #fb=np.sin(np.pi/(2*N)*(np.arange(int(1.5*N))+0.5))
    fb=np.loadtxt("MDCTcoeff.txt") #Coeff. from optimization
    print("fb=", fb)
    #input test signal, ramp:
    x=np.arange(64)
    plt.plot(x)
    plt.title('Input_Signal')
    plt.xlabel('Sample')
    plt.show()
    y=MDCTanafb(x,N,fb); print("y=\n", y)
    plt.imshow(np.abs(y))
    plt.title('MDCT_Subbands')
    plt.xlabel('Block_No.')
    plt.ylabel('Subband_No.')
    plt.show()
    xr=MDCTsynfb(y,fb)
    plt.plot(xr)
    plt.title('Reconstructed_Signal')
    plt.xlabel('Sample')
    plt.show()
    y=np.zeros((4,16))
    y[0,0]=1
    xr=MDCTsynfb(y,fb)
    plt.plot(xr[0:3*N])
```

plt.title('Impulse_Response_of_Modulated_Synthesis_Subband_0')
plt.xlabel('Sample')

and execute it with

python MDCTfb.py

1.13 Further Extending the Impulse Response Length

We can now use this framework to further extend the length of the filters of our filter bank ([20], see also [21]). For longer filters we obtain polynomials of higher order in our polyphase matrices, since we need more blocks of size N for the impulse response. If we look at our analysis folding matrix $\mathbf{F_a}(z)$ for our modulation function, it will take the following form for longer filters:

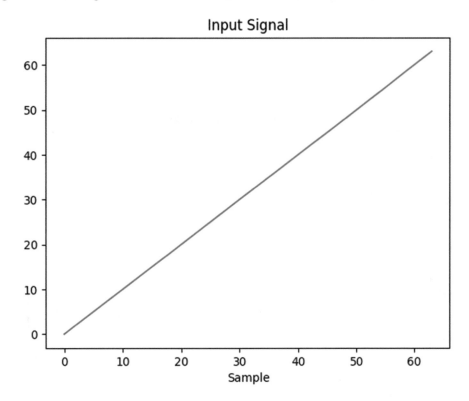

Figure 1.21: The ramp testing signal for the analysis filter bank input.

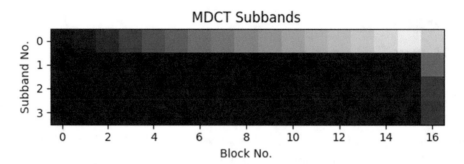

Figure 1.22: The resulting subband signals after analysis filtering of the ramp testing signal. The different colours symbolize different signal values. Observe that most of the energy appears in subband 0.

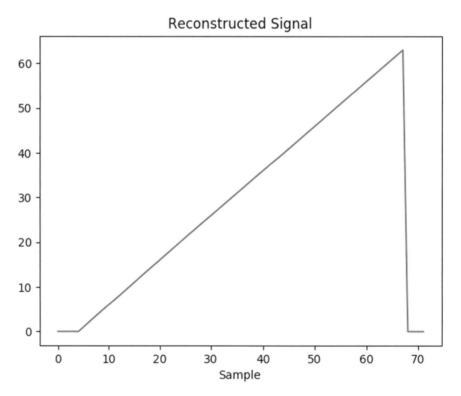

Figure 1.23: The reconstructed ramp signal. Any corruption would be easy to spot here. Observe the delay of 1 block of N samples (the result of making the inverse delay matrix causal).

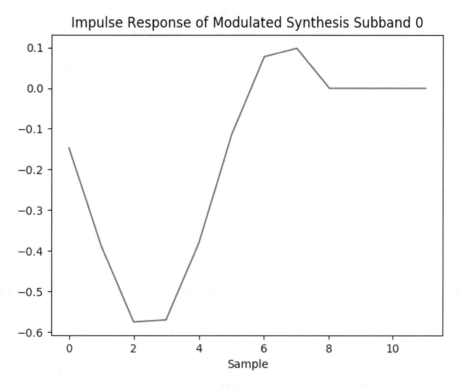

Figure 1.24: The impulse response of synthesis filter 0, obtained by using all zeros as subbands signals input for the synthesis filter bank, except a single "1" in subband 0. Observe that this is indeed a continuous smooth function. It looks different from the prototype function because it is the prototype multiplied by the cosine modulation function for $k = 0$.

$$\mathbf{F_a}(z) = \begin{bmatrix} & & H_{2N-1}^{\downarrow 2N}(-z^2)z^{-1}, & H_{N-1}^{\downarrow 2N}(-z^2) & & \\ & & \ddots & & \ddots & \\ H_{N+N/2}^{\downarrow 2N}(-z^2)z^{-1} & & & & & H_{N/2}^{\downarrow 2N}(-z^2) \\ H_{N+N/2-1}^{\downarrow 2N}(-z^2)z^{-1} & & & & & -H_{N/2-1}^{\downarrow 2N}(-z^2) \\ & \ddots & & & \ddots & \\ & & H_{N}^{\downarrow 2N}(-z^2)z^{-1}, & -H_0^{\downarrow 2N}(-z^2) & & \end{bmatrix}$$

with

$$H_n^{\downarrow 2N}(z) = \sum_{m=0}^{\infty} h(m2N + n)z^{-m}. \tag{1.59}$$

Hence if we have a given folding matrix F_a, we can read out the baseband impulse response $h(n)$ by simply comparing it with this general formulation. Similar for the synthesis, the general form of the synthesis folding matrix $F_s(z)$ for our modulation function is

$$\mathbf{F_s}(z) = \begin{bmatrix} & & H'^{\downarrow 2N}_{N/2-1}(-z^2), & H'^{\downarrow 2N}_{N/2}(-z^2) & & \\ & \cdot\cdot & & & & \cdot\cdot \\ & H'^{\downarrow 2N}_0(-z^2) & & & & H'^{\downarrow 2N}_{N-1}(-z^2) \\ & H'^{\downarrow 2N}_N(-z^2)z^{-1} & & & & -H'^{\downarrow 2N}_{2N-1}(-z^2)z^{-1} \\ & \cdot\cdot & & & & \cdot\cdot \\ & & H'^{\downarrow 2N}_{N+N/2-1}(-z^2)z^{-1}, & -H'^{\downarrow 2N}_{N+N/2}(-z^2)z^{-1} & & \end{bmatrix}$$

with

$$H'^{\downarrow 2N}_n(z) = \sum_{m=0}^{\infty} h'(m2N+n)z^{-m}. \tag{1.60}$$

The indexing of the sum goes to infinity to accommodate the synthesis filters that can be IIR. This we again could use to read out our baseband impulse response, now for the synthesis side.

An alternative way to obtain the baseband impulse responses is to multiply our folding matrix with the DCT type 4 to obtain the filter banks Polyphase matrix, read out the lowest subband filter from the corresponding column (or row), and divide it by the modulation function for the lowest subband.

Using this matrix structure, we can again calculate the general matrix inverse by looking at the 2x2 submatrices. The submatrices of the analysis folding matrix are $\mathbf{F_a}(z)$

$$\begin{bmatrix} H^{\downarrow 2N}_{2N-1-n}(z), & H^{\downarrow 2N}_{N-1-n}(z) \\ H^{\downarrow 2N}_{N+n}(z), & -H^{\downarrow 2N}_n(z) \end{bmatrix}$$

For the synthesis folding matrix $\mathbf{F_s}(z)$ the corresponding 2x2 submatrix is

$$\begin{bmatrix} H'^{\downarrow 2N}_n(z), & H'^{\downarrow 2N}_{N-1-n}(z) \\ H'^{\downarrow 2N}_{N+n}(z) & -H'^{\downarrow 2N}_{2N-1-n}(z) \end{bmatrix}$$

Then for $\mathbf{F_s}(z) = \mathbf{F_a}^{-1}(z)$ we obtain

$$\begin{bmatrix} H'^{\downarrow 2N}_n(z) & H'^{\downarrow 2N}_{N-1-n}(z) \\ H'^{\downarrow 2N}_{N+n}(z) & -H'^{\downarrow 2N}_{2N-1-n}(z) \end{bmatrix} =$$

$$= \frac{1}{H^{\downarrow 2N}_{2N-1-n}(z)H^{\downarrow 2N}_n(z) + H^{\downarrow 2N}_{N-1-n}(z)H^{\downarrow 2N}_{N+n}(z)} \begin{bmatrix} H^{\downarrow 2N}_n(z) & H^{\downarrow 2N}_{N-1-n}(z) \\ H^{\downarrow 2N}_{N+n}(z) & -H^{\downarrow 2N}_{2N-1-n}(z) \end{bmatrix}$$

In this more general case, we can now again see that if the negative determinant of the 2x2 submatrices is ± 1 (or in general a delay), then the **analysis and synthesis prototypes are identical!**

$$h'(n) = \pm h(n) \tag{1.61}$$

Now assume (as for the MDCT case) that we have an orthogonal filter bank with **para-unitary polyphase matrices**, which means

$$\mathbf{P_a}^{-1}(z) = \mathbf{P_a}^T(z^{-1})$$

1.13 Further Extending the Impulse Response Length

This results in

$$
\frac{1}{H_{2N-1-n}^{\downarrow 2N}(z)H_n^{\downarrow 2N}(z) + H_{N-1-n}^{\downarrow 2N}(z)H_{N+n}^{\downarrow 2N}(z)}
\begin{bmatrix} H_n^{\downarrow 2N}(z) & H_{N-1-n}^{\downarrow 2N}(z) \\ H_{N+n}^{\downarrow 2N}(z) & -H_{2N-1-n}^{\downarrow 2N}(z) \end{bmatrix} =
$$

$$
= \begin{bmatrix} H_{2N-1-n}^{\downarrow 2N}(z^{-1}) & H_{N+n}^{\downarrow 2N}(z^{-1}) \\ H_{N-1-n}^{\downarrow 2N}(z^{-1}) & -H_n^{\downarrow 2N}(z^{-1}) \end{bmatrix}
$$

The argument z^{-1} results in a time reversal of their polyphase components. Now we can see in this more general case that we obtain a para-unitary polyphase matrix if the negative determinant $P_{2N-1-n}(z)P_n(z) + P_{N-1-n}(z)P_{N+n}(z)$ is one (which again leads to the baseband prototype impulse responses being identical between analysis and synthesis), and if the baseband prototype impulse response is

$$
h(n) = h(LN - 1 - n) \tag{1.62}
$$

meaning it is **symmetric around its centre**, also in this general case.

1.13.1 Low Delay Filter Banks

These Low Delay Filter Banks are used, e.g., in the MPEG-4 Enhanced Low Delay AAC coder [22, 23]. We still need to think about the invertibility of the resulting polyphase matrices. Hence our approach is to construct the polyphase matrices of our filter bank as a product of matrices with lower order polynomials. The product of matrices then leads to polynomials of higher order in the resulting polyphase matrix, and hence longer filters. To obtain the inverse of the resulting polyphase matrix, we only need to obtain the inverse of the simpler matrices with lower order polynomials, which is a simpler task. We can also construct those simpler matrices such that their inverse again has finite polynomials, leading to FIR filters. The inverse of those simpler matrices might need a delay to make the inverse matrix causal, to avoid positive exponents of z. This delay to make the inverses causal then leads to a resulting system delay of the constructed filter bank. To keep the system delay low, it would be very helpful to have a matrix, whose inverse does not need any delay to make it causal, because it is already causal [24].

The following matrix fulfils this condition, which we call **Zero-Delay Matrix** for that reason,

$$
\mathbf{G}_i(z) = \begin{bmatrix} z^{-1}g_0^i & & & & 0 & & & 1 \\ & \ddots & & & & & \iddots & \\ & & & z^{-1}g_{N/2-1}^i & 1 & & & \\ & & & 1 & 0 & & & \\ & \iddots & & & & & \ddots & \\ 1 & & & & 0 & & & 0 \end{bmatrix} \tag{1.63}
$$

Since we can have more than one Zero-Delay matrix, and to distinguish it from the synthesis polyphase matrix, we give it an index, i. Its inverse is causal, hence it does not need any delay to make it causal,

$$
\mathbf{G}_i^{-1}(z) = \begin{bmatrix} 0 & & & & & & 1 \\ & \ddots & & & & \iddots & \\ & & 0 & 1 & & & \\ & & 1 & -z^{-1}g_{N/2-1}^i & & & \\ & \iddots & & & & \ddots & \\ 1 & & & & & & -z^{-1}g_0^i \end{bmatrix}
$$

Observe that we obtain the inverse by simply flipping the elements on the anti-diagonal to its other half and changing their sign. Hence the exponents of z stay negative and causality is maintained.

Another Zero-Delay Matrix type is

$$
\mathbf{E}_i(z) = \begin{bmatrix} 0 & & 0 & & & 1 \\ & \ddots & & & \iddots & \\ & & 0 & 1 & & \\ & & 1 & z^{-1}e_0^i & & \\ & \iddots & & & \ddots & \\ 1 & & 0 & & & z^{-1}e_{N/2-1}^i \end{bmatrix} \tag{1.64}
$$

which has the non-zero elements on the other half on the matrix diagonal, with the same property of a causal inverse.

If we multiply this matrix with another polyphase matrix, we obtain a polyphase matrix with higher order polynomials and hence filters with increased length, without increasing the system delay of the resulting filter bank. In a pole-zero diagram of the filters of the filter bank this can be seen as adding zeros inside the unit circle. Since the matrix is easily invertible, the coefficients l_i don't affect the perfect reconstruction property, but the resulting impulse and frequency responses of the filters. To obtain desired frequency responses, we can apply numerical optimization to find suitable coefficients l_i. Note that the multiplication of our Folding matrix with this Zero-Delay matrix keeps the highest exponents of z^{-1} on the same side. This is associated with a modulation function which ends on the same phase after the extension of the length, which explains the implementation with the reverse time index of the cosine modulation in the function Fa2h.py.

It might not always be desirable to have the lowest delay. For instance, often better stop band attenuation can be obtained with a medium delay. To have the freedom to also design filter banks with different delays, we use a matrix which looks very similar to the Zero-Delay matrix, but has the opposite delay property, a maximum delay. Hence we call it a **Maximum-Delay Matrix**. We obtain it by replacing each z^{-1} by z in the zero delay matrix, and then multiplying the matrix with z^{-1} to make it causal,

$$
\mathbf{H2}_i(z) = z^{-1} \cdot \mathbf{G}_i(z^{-1}) =
$$

$$
= \begin{bmatrix} h2^i_0 & & 0 & & z^{-1} \\ & \ddots & & & \cdot^{\cdot^{\cdot}} \\ & & h2^i_{N/2-1} & z^{-1} & \\ & & z^{-1} & 0 & \\ & \cdot^{\cdot^{\cdot}} & & & \ddots \\ z^{-1} & & 0 & & 0 \end{bmatrix}
$$

Hence its inverse with the delay to make it causal is

$$
\mathbf{H2}_i^{-1}(z) \cdot z^{-2} = z^{-1} \cdot \mathbf{G}_i^{-1}(z^{-1}) =
$$

$$
= \begin{bmatrix} 0 & & 0 & & z^{-1} \\ & \ddots & & & \cdot^{\cdot^{\cdot}} \\ & & 0 & z^{-1} & \\ & & z^{-1} & -h2^i_{N/2-1} & \\ & \cdot^{\cdot^{\cdot}} & & & \ddots \\ z^{-1} & & 0 & & -h2^i_0 \end{bmatrix}
$$

Observe that we need a factor of z^{-1} to make the inverse causal, which is the result of the two factors of z^{-1} for $\mathbf{G}_i(z^{-1})$ and $\mathbf{G}_i^{-1}(z^{-1})$. This means that using this matrix for the construction of our polyphase matrix leads to an increase of the system delay by 2 blocks of N samples. It can also be seen as adding zeros in the pole/zero plot of the filters outside the unit circle.

Another type of Maximum-Delay matrix with the same properties is

$$
\mathbf{H1}_i(z) = z^{-1} \cdot \mathbf{E}_i(z^{-1})
$$

Now that we have building blocks to construct a polyphase matrix with desired length and delay properties, the approach is to start with an MDCT like polyphase matrix and extend it by multiplying it with the desired number of Low Delay and High Delay matrices. In this way we can obtain filter banks with a system delay lower than for filter banks with para-unitary polyphase matrices. For that reason we call them Low Delay Filter Banks. We now use the indices i and j for the different low delay and high delay matrices, where each of these matrices has its own parameter set. We obtain their analysis polyphase matrix as

$$
\mathbf{H}(z) = \mathbf{F} \cdot \mathbf{D}(z) \cdot \prod_{i=0}^{\mu-1} \mathbf{G}_i(z) \cdot \prod_{j=0}^{\nu-1} \mathbf{H}_j(z) \cdot \mathbf{T} \tag{1.65}
$$

where $\mathbf{H}_j(z)$ is either $\mathbf{H1}_j(z)$ if μ is even, or $\mathbf{H2}_j(z)$, if μ is odd. The system delay of the resulting filter bank consists of the delays needed to make the inverses of the matrices $\mathbf{D}(z)$ and $\mathbf{H}_i(z)$ causal (N and $2N$ samples resp.), plus the blocking delay, which results from assembling the signals into blocks of size N, which is $N-1$ samples. Hence the total system delay will be

$$
n_d = N - 1 + N + \nu \cdot 2N.
$$

For the resulting filter length L, each low delay matrix and high delay matrix adds N taps to the filter length. Observe that this is valid for μ even. For odd μ we need to use a low delay matrix which has its coefficients on the other half of the anti-diagonal to obtain the same filter length. We start with an MDCT like polyphase matrix, which leads to filters of length $2N$. Hence the resulting filter length for our low delay filter bank will be

$$L = 2N + (\mu + \nu) \cdot N.$$

The synthesis polyphase matrix for perfect reconstruction is then simply the inverse with the suitable delays to make it causal,

$$\mathbf{G}(z) = \mathbf{T}^{-1} \prod_{j=\nu-1}^{0} \left(\mathbf{H}_j^{-1}(z) \cdot z^{-2} \right) \cdot \prod_{i=\mu-1}^{0} \mathbf{G}_i^{-1}(z) \cdot \mathbf{D}^{-1}(z) \cdot z^{-1} \cdot \mathbf{F}^{-1} \quad (1.66)$$

To obtain a desired frequency response for the filters, we use numerical optimization for the coefficients of our matrices to obtain a resulting baseband prototype filter which features the desired properties (good stopband attenuation, desired transition band,...). Observe that perfect reconstruction is not part of the optimization, because it is structurally guaranteed.

Hence the approach for the design of low delay filter banks would be to first determine the necessary total number of Zero-Delay and Maximum-Delay matrices according to the desired filter length. The desired delay then gives the number of Maximum-Delay matrices, and the desired frequency response is the goal for a numerical optimization routine.

1.13.2 Python Example for Low Delay Filter Banks

To optimize a low delay filter bank, we first need to implement the Zero-Delay matrix $\mathbf{G_i}(z)$ as a function in Python,

```python
def Gmatrix(g):
    '''produces a zero delay matrix G(z), which has delays z^-1 multiplied
    with the g coefficients on the upper half of its diagonal. g is a row vector
    with N/2 coefficients.
    In a 3D polynomial representation (exponents of z^-1 are in the third dimension)
    N is number of subbands and size of the polynomial matrix (NxN)
    N is even'''
    import numpy as np

    N = max(np.shape(g))*2;
    G = np.zeros((N,N,2));
    G[:,:,0] = np.fliplr(np.eye(N))
    G[:,:,1] = np.diag(np.concatenate((g,np.zeros(int(N/2)))))
    return G
```

Our analysis folding matrix here is $\mathbf{F_a}(z) = \mathbf{F} \cdot \mathbf{D}(z) \cdot \mathbf{G}_0(z)$, according to Eq. (1.65) with $\mu = 0$ and $\nu = 1$. We put all unknown coefficients into 1-d vector $\mathbf{x}$. Hence we can construct our error function for the optimization, and store it in the file "optimfuncLDFB.py",

```
import numpy as np
import scipy as sp
import scipy.signal as sig
from symFmatrix import symFmatrix
from polmatmult import polmatmult
from Dmatrix import Dmatrix
from Fa2h import Fa2h
from Gmatrix import Gmatrix


def optimfuncLDFB(x,N):
    '''function for optimizing an MDCT type filter bank.
    x: unknown matrix coefficients, N: Number of subbands.
    '''
    #Analysis folding matrix:
    Fa = symFmatrix(x[0:int(1.5*N)])
    Faz = polmatmult(Fa,Dmatrix(N))
    Faz = polmatmult(Faz,Gmatrix(x[int(1.5*N):(2*N)]))
    #baseband prototype function h:
    h = Fa2h(Faz)
    #'Fa2h is returning 2D array. Squeeze to make it 1D'
    #h= np.squeeze(h)

    #Frequency response of the the baseband prototype function at 1024 frequency sampling
        points
    #between 0 and Nyquist:
    w,H = sig.freqz(h,1,1024)
    #desired frequency response
    #Limit of desired pass band (passband is between -pb and +pb, hence ../2)
    pb = int(1024/N/2.0)
    #Ideal desired frequency response:
    Hdes = np.concatenate((np.ones(pb),np.zeros(1024-pb)))
    #transition band width to allow filter transition from pass band to stop band:
    tb = int(np.round(1.0*pb))
    #Weights for differently weighting errors in pass band, transition band and stop band:
    weights = np.concatenate((1.0*np.ones(pb), np.zeros(tb),1000*np.ones(1024-pb-tb)))
    #Resulting total error number as the sum of all weighted errors:
    err = np.sum(np.abs(H-Hdes)*weights)
    return err
```

Here we use the Python optimization package "scipy.optimize.differential_evolution" as the optimization algorithm, because it finds better results. Again, we can put the minimization into the "main" routine of the file "optimfuncLDFB.py",

```python
if __name__ == '__main__': #run the optimization
  import scipy as sp
  import scipy.signal
  import matplotlib.pyplot as plt


  N=4 #number of subbands
  s=2*N
  bounds=[(-14,14)]*s
  xmin = sp.optimize.differential_evolution(optimfuncLDFB, bounds, args=(N,), disp=True)
  print("error_after_optim.=",xmin.fun)
  print("optimized_coefficients=",xmin.x)
  np.savetxt("LDFBcoeff.txt", xmin.x)
  x=xmin.x;
  #Baseband Impulse Response:
  Fa = symFmatrix(x[0:int(1.5*N)])
  #print("Fa=", Fa[:,:,0])
  Faz = polmatmult(Fa,Dmatrix(N))
  Faz = polmatmult(Faz,Gmatrix(x[int(1.5*N):(2*N)]))
  h = Fa2h(Faz)
  plt.plot(h)
  plt.xlabel('Sample')
  plt.ylabel('Value')
  plt.title('Baseband_Impulse_Response_of_the_Low_Delay_Filter_Bank')
  #Magnitude Response:
  w,H=sp.signal.freqz(h)
  plt.figure()
  plt.plot(w,20*np.log10(abs(H)))
  plt.axis([0, 3.14, -60,20])
  plt.xlabel('Normalized_Frequency')
  plt.ylabel('Magnitude_(dB)')
  plt.title('Mag._Frequency_Response_of_the_Low_Delay_Filter_Bank')
  plt.show()
```

We start the minimization with

```
python optimfuncLDFB.py
```

The resulting coefficients are stored in the text file "LDFBcoeff.txt". Figure 1.25 shows our resulting baseband impulse response. Observe that it now has a length of 12 samples (instead of 8 for the MDCT), and that it became non-symmetric. The resulting magnitude of the frequency response can be seen in Fig. 1.26.

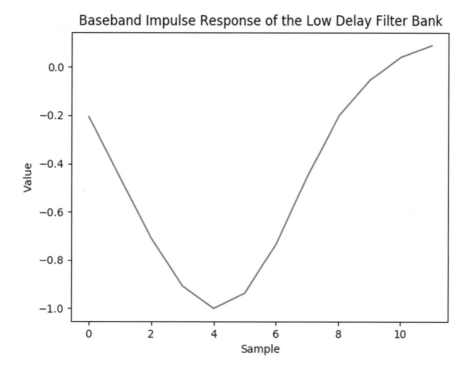

Figure 1.25: The baseband prototype impulse response for the low delay filter bank, resulting from our optimization.

Observe that in comparison with the MDCT optimization, we obtain a narrower transition band, with the stopband starting at lower frequencies, and that the first sidelobe at about frequency 1.1 now has better than -30 dB attenuation relative to the passband (compared to about -25 dB for the MDCT). This means we obtain an improved filter bank without increasing the system delay of our filter bank, the delay is still the same as the MDCT filter bank (at $2N - 1$, including the blocking delay of $N - 1$ samples, or a delay of 1 block of N samples without the blocking delay).

Low Delay Filter Bank Implementation

Now we have all the functions and the coefficients to implement the low delay filter bank we optimized in the previous section. For the synthesis filter bank we still need the inverse of the Zero-Delay matrix, for which we use the function Ginvmatrix,

```
def Ginvmatrix(g):
    '''produces the inverse of the zero delay matrix G(z). It has delays z^-1 multiplied
    with the reverse ordered neg. g coefficients on the lower half of its diagonal. g is a 1-d
        vector with N/2 coefficients.
    In a 3D polynomial representation (exponents of z^-1 are in the third dimension).
    N is number of subbands and size of the polynomial matrix (NxN)
    N is even'''
    import numpy as np
    N = max(np.shape(g))*2;
    G = np.zeros((N,N,2));
```

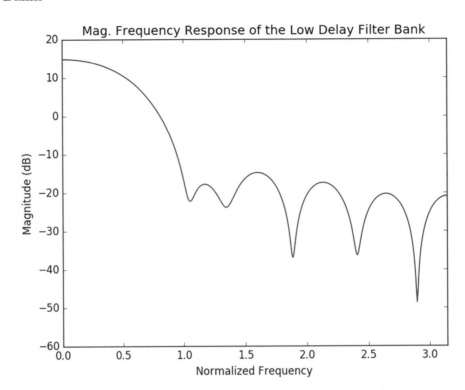

Figure 1.26: The baseband prototype frequency response of our Low Delay filter bank resulting from our optimization.

G[:,:,0] = np.fliplr(np.eye(N))
G[:,:,1] = np.diag(np.concatenate((np.zeros(**int**(N/2)), −np.flipud(g))))
return G

The low delay analysis filter bank is then implemented with the following function, stored in file

def Ginvmatrix(g):
  '''*produces the inverse of the zero delay matrix G(z). It has delays z^-1 multiplied with the reverse ordered neg. g coefficients on the lower half of its diagonal. g is a $1-d$ vector with N/2 coefficients.*
  In a 3D polynomial representation (exponents of z^-1 are in the third dimension).
  N is number of subbands and size of the polynomial matrix (NxN)
  N is even'''
  import numpy as np
  N = **max**(np.shape(g))*2;
  G = np.zeros((N,N,2));
  G[:,:,0] = np.fliplr(np.eye(N))
  G[:,:,1] = np.diag(np.concatenate((np.zeros(**int**(N/2)), −np.flipud(g))))
  return G

Now the analysis low delay filter bank can be implemented with the following function, which we store in file "LDFB.py", as we do with the further functions.

```
from __future__ import print_function
from symFmatrix import symFmatrix
from polmatmult import polmatmult
from Dmatrix import Dmatrix
from Gmatrix import Gmatrix
from DCT4 import *
from x2polyphase import *

def LDFBana(x,N,fb):
    #Low Delay analysis filter bank.
    #Arguments: x: input signal, e.g. audio signal, a 1-dim. array
    #N: number of subbands
    #fb: coefficients for the MDCT filter bank, for the F matrix, np.array with 1.5*N
        coefficients.
    #returns y, consisting of blocks of subband in in a 2-d array of shape (N,# of blocks)

    Fa=symFmatrix(fb[0:int(1.5*N)])
    print("Fa.shape=",Fa.shape)
    D=Dmatrix(N)
    G=Gmatrix(fb[int(1.5*N):(2*N)])
    y=x2polyphase(x,N)
    print("y[:,:,0]=", y[:,:,0])
    y=polmatmult(y,Fa)
    y=polmatmult(y,D)
    y=polmatmult(y,G)
    y=DCT4(y)
    #strip first dimension:
    y=y[0,:,:]
    return y
```

The synthesis low delay filter bank is the following function:

```
from Dinvmatrix import Dinvmatrix
from Ginvmatrix import *
from polyphase2x import *

def LDFBsyn(y,fb):
    #Low Delay synthesis filter bank.
    #Arguments: y: 2-d array of blocks of subbands, of shape (N, # of blokcs)
    #returns xr, the reconstructed signal, a 1-d array.

    Fa=symFmatrix(fb[0:int(1.5*N)])
    #invert Fa matrix for synthesis after removing last dim:
    Fs=np.linalg.inv(Fa[:,:,0])
    #add again last dimension for function polmatmult:
    Fs=np.expand_dims(Fs, axis=-1)
    Ginv=Ginvmatrix(fb[int(1.5*N):(2*N)])
    Dinv=Dinvmatrix(N)
    #Display the synthesis folding matrix Fs(z):
```

```
    Fsz=polmatmult(polmatmult(Ginv,Dinv),Fs)
    #add first dimension to y for polmatmult:
    y=np.expand_dims(y,axis=0)
    xp=DCT4(y)
    xp=polmatmult(xp,Ginv)
    xp=polmatmult(xp,Dinv)
    xp=polmatmult(xp,Fs)
    xr=polyphase2x(xp)
    return xr
```

Then we add the main routine for testing, in the same way as we did for testing the MDCT filter bank (Figs. 1.27, 1.28, 1.29, and 1.30).

```
#Testing:
if __name__ == '__main__':
    import numpy as np
    import matplotlib.pyplot as plt

    #Number of subbands:
    N=4

    #D=Dmatrix(N)
    #Dinv=Dinvmatrix(N)
    #Filter bank coefficients, 1.5*N of sine window:
    #fb=np.sin(np.pi/(2*N)*(np.arange(int(1.5*N))+0.5))
    fb=np.loadtxt("LDFBcoeff.txt")
    print("fb=", fb)
    #input test signal, ramp:
    x=np.arange(64)
    plt.plot(x)
    plt.title('Input_Signal')
    plt.xlabel('Sample')
    plt.show()
    y=LDFBana(x,N,fb)
    plt.imshow(np.abs(y))
    plt.title('LDFB_Subbands')
    plt.xlabel('Block_No.')
    plt.ylabel('Subband_No.')
    plt.show()
    xr=LDFBsyn(y,fb)
    plt.plot(xr)
    plt.title('Reconstructed_Signal')
    plt.xlabel('Sample')
    plt.show()
    #Input to the synthesis filter bank: unit pulse in lowest subband
    #to see its impulse response:
    y=np.zeros((4,16))
```

```
y[0,0]=1
xr=LDFBsyn(y,fb)
plt.plot(xr[0:4*N])
plt.title('Impulse_Response_of_Modulated_Synthesis_Subband_0')
plt.xlabel('Sample')
plt.show()
```

The next figures show an example of a low delay filter bank with a higher number of subbands, for $N = 128$, and comparison with the MDCT with a standard delay of $2N - 1 = 255$ samples. Figure 1.31 shows the two impulse responses, and Fig. 1.32 the resulting magnitude of the frequency responses. Observe that the Low Delay filter bank has a similar passband and transition band as the MDCT, but about 20 dB more stopband attenuation! It again shows that low delay filter banks can be used to improve the frequency response (to increase the stopband attenuation) of a filter bank by increasing the length of the impulse response of its filters while keeping the same system delay of the filter bank.

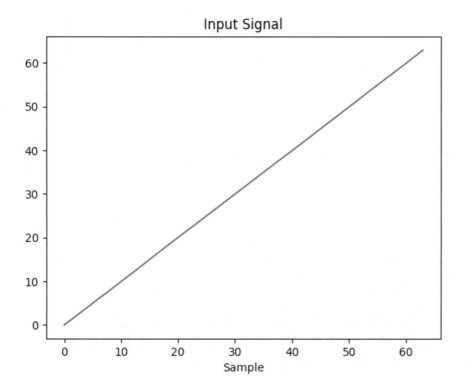

Figure 1.27: The ramp testing signal for the analysis filter bank input.

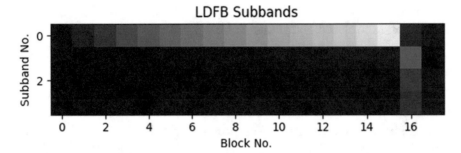

Figure 1.28: The resulting subband signals after analysis filtering of the ramp testing signal. The different colours symbolize different signal values. Observe that most of the energy appears in subband 0. Because of the longer filters we have 1 more block of subbands than for the MDCT.

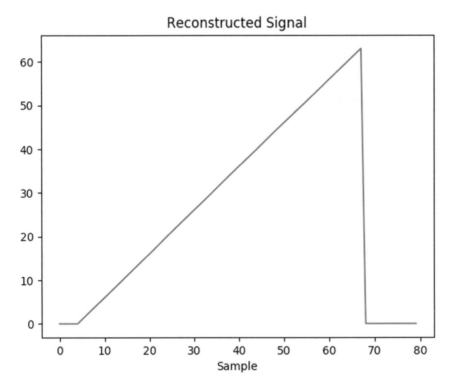

Figure 1.29: The reconstructed ramp signal. Any corruption would be easy to spot here. Observe the delay of 1 block of N samples, the same as for the MDCT, even though we have longer filters here!.

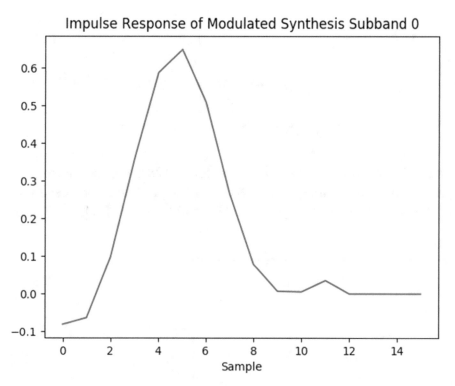

Figure 1.30: The impulse response of synthesis filter 0, obtained by using all zeros as subbands signals input for the synthesis filter bank, except a single "1" in subband 0. Observe that this is indeed a continuous smooth function, and longer than for the MDCT.

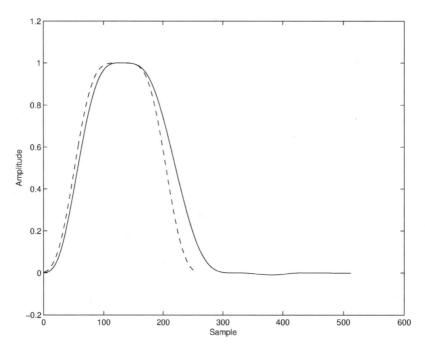

Figure 1.31: The baseband prototype impulse response for a 128 subband optimized MDCT filter bank (dashed line), compared to a 128 subband optimized low delay filter bank with the same system delay (solid line), from [25].

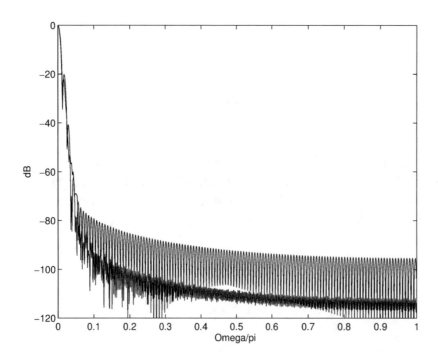

Figure 1.32: The baseband prototype frequency response for the 128 subband optimized MDCT filter bank, compared to the 128 subband optimized low delay filter bank with the same system delay (lower line), from [25].

1.14 Pseudo Quadrature Mirror Filter Banks (PQMF)

Pseudo Quadrature Mirror or PQMF filter banks are used in MPEG1/2 Layer 1 and 2, as 32 subband filter bank [3], and also in the newer MPEG 4 standards for implementing spatial audio coding [26] and audio bandwidth extension [27]. PQMF filter banks are defined with their analysis filter impulse responses as

$$h_k(n) = h(n) \cdot \cos\left(\frac{\pi}{N} \cdot (k + 0.5) \cdot (n - \frac{LN}{2} + \frac{1}{2}) + \phi_k\right) \qquad (1.67)$$

where L is the length of the filters in blocks of size N. The phase offset ϕ_k is defined as [12, 28–30],

$$\phi_{k+1} - \phi_k = (2r + 1) \cdot \frac{\pi}{2}$$

with some integer r. The PQMF is an orthogonal filter bank, hence the baseband prototype $h(n)$ is symmetric around its centre and the synthesis filters are obtained by time-reversing the analysis filters,

$$g_k(n) = h_k(LN - 1 - n)$$

This is again a cosine modulated filter bank. By inspection, its modulation function is quite similar to the modulation function of the MDCT and low delay filter banks of Eq. (1.46) (without the normalization factor),

$$h_k(n) = h(n) \cdot \cos\left(\frac{\pi}{N}\left(k + \frac{1}{2}\right)\left((LN - 1 - n) - \frac{N}{2} + \frac{1}{2}\right)\right)$$

To obtain a positive time index n we can reformulate it by switching the sign of the cosine argument and rearrange it,

$$h_k(n) = h(n) \cdot \cos\left(\frac{\pi}{N}\left(k + \frac{1}{2}\right)\left(n - LN + \frac{N}{2} + \frac{1}{2}\right)\right)$$

Indeed, it is interesting to observe that this modulation function becomes identical to the MDCT modulation function for specific phase functions ϕ_k and certain length L, except for a sign change. But a sign change of a filter does not really change the filter bank, because a sign change of an analysis filter is accompanied by the same sign change for the synthesis filter, which then cancels the sign changes. Our specific phase function is [29]

$$\phi_k = (-1)^{k+1} \cdot \frac{\pi}{4}$$

and the length fulfills

$$L \mod 4 \equiv \pm 2$$

We can now rewrite Eq. (1.67) as

$$h_k(n) = h(n) \cdot \cos\left(\frac{\pi}{N} \cdot (k+0.5) \cdot (n - LN + \frac{N}{2} + \frac{1}{2}) + \frac{\pi}{N} \cdot (k + \frac{1}{2}) \cdot \frac{(L-1)N}{2} + \phi_k\right)$$

(1.68)

Observe that the first part of the cos argument is already identical to the argument of the MDCT modulation function. To show that the entire modulation function is identical to the MDCT modulation function, up to a sign change, we just have to show that the right half of the argument is an integer multiple of π,

$$\frac{\pi}{N} \cdot (k+0.5) \cdot \frac{(L-1)N}{2} + \phi_k =$$

$$= \pi \cdot (k+0.5) \cdot \frac{(L-1)}{2} + (-1)^{k+1} \cdot \frac{\pi}{4} =$$

$$= \pi\left(\frac{kL}{2} + \frac{L}{4} - \frac{k}{2} - \frac{1}{4} + \frac{(-1)^{k+1}}{4}\right)$$

To show that this expression is an integer multiple of π, we just have to determine if the fractional part of the expression within the parentheses is indeed zero. We do that first for k even,

$$\left(\frac{1}{2} - \frac{1}{4} - \frac{1}{4}\right) = 0$$

we see that it indeed results to zero. Now for k odd,

$$\left(\frac{1}{2} - \frac{1}{2} - \frac{1}{4} + \frac{1}{4}\right) = 0$$

so we see that it is indeed zero also in this case. This means the modulation function is indeed identical, up to a sign change, to the MDCT and low delay filter bank modulation function (1.46)!

Now we can use our matrix formulation for cosine modulated filter banks. The QMF analysis folding matrix is

$$\mathbf{F_a} = \begin{bmatrix} & & H_{2N-1}^{\downarrow 2N}(-z^2)z^{-1} & H_{N-1}^{\downarrow 2N}(-z^2) & & \\ & \iddots & & & \ddots & \\ H_{N+N/2}^{\downarrow 2N}(-z^2)z^{-1} & & & & & H_{N/2}^{\downarrow 2N}(-z^2) \\ H_{N+N/2-1}^{\downarrow 2N}(-z^2)z^{-1} & & & & & -H_{N/2-1}^{\downarrow 2N}(-z^2) \\ & \ddots & & & \iddots & \\ & & H_N^{\downarrow 2N}(-z^2)z^{-1} & -H_0^{\downarrow 2N}(-z^2) & & \end{bmatrix}$$

with

$$H_k^{\downarrow 2N}(z) = \sum_{m=0}^{\infty} h(m2N+k)z^{-m}.$$

(1.69)

and for the synthesis it is

$$\mathbf{F_s} = \begin{bmatrix} & & H'^{\downarrow 2N}_{N/2-1}(-z^2) & H'^{\downarrow 2N}_{N/2}(-z^2) & & & \\ & \ddots & & & & \ddots & \\ H'^{\downarrow 2N}_0(-z^2) & & & & & & H'^{\downarrow 2N}_{N-1}(-z^2) \\ H'^{\downarrow 2N}_N(-z^2)z^{-1} & & & & & & -H'^{\downarrow 2N}_{2N-1}(-z^2)z^{-1} \\ & \ddots & & & & \ddots & \\ & & H'^{\downarrow 2N}_{N+N/2-1}(-z^2)z^{-1} & -H'^{\downarrow 2N}_{N+N/2}(-z^2)z^{-1} & & & \end{bmatrix}$$

with

$$H'^{\downarrow 2N}_k(z) = \sum_{m=0}^{\infty} h'(m2N+k)z^{-m}. \tag{1.70}$$

For a near-perfect reconstruction filter bank we need the **off-diagonal elements** of the product $\mathbf{F_a}(z) \cdot \mathbf{F_s}(z)$ to **disappear**. These off-diagonal elements are

$$z^{-1} \cdot (H^{\downarrow 2N}_n(-z^2) \cdot H'^{\downarrow 2N}_{N+n}(-z^2) - H^{\downarrow 2N}_{N+n}(-z^2) \cdot H'^{\downarrow 2N}_n(-z^2))$$

and

$$z^{-1} \cdot (H^{\downarrow 2N}_{N-1-n}(-z^2) \cdot H'^{\downarrow 2N}_{2N-1-n}(-z^2) - H^{\downarrow 2N}_{2N-1-n}(-z^2) \cdot H'^{\downarrow 2N}_{N-1-n}(-z^2)).$$

We see that they become indeed zero if we choose $H'^{\downarrow 2N}_n(z) = H^{\downarrow 2N}_n(z)$, which is the case if the baseband prototype impulse responses are identical for analysis and synthesis,

$$H(z) = H'(z). \tag{1.71}$$

The **diagonal elements** of the product then have the form

$$\pm delay = z^{-1}\left(H^{\downarrow 2N}_n(-z^2) \cdot H^{\downarrow 2N}_{2N-1-n}(-z^2) + H^{\downarrow 2N}_{N+n}(-z^2) \cdot H^{\downarrow 2N}_{N-1-n}(-z^2)\right) \tag{1.72}$$

for $n = 0, \ldots, N/2 - 1$, which is exactly the **negative determinant** of the 2×2 submatrices, which needs to be approximately 1 or a delay for near-perfect reconstruction. Here we have a possible sign change ($\pm$) in front of the delay, because it is more general and we can always switch the sign back after reconstruction. All which is left for our PQMF design for near-perfect reconstruction is now to ensure that this condition on the determinant is indeed true, at least approximately.

We can interpret (1.72) as a polyphase representation of a filter with a downsampling factor of $N = 2$ (because of the 2 polyphase components), followed by upsampling (because all arguments are in z^2), as shown in Fig. 1.33. The downsampling leads to a polyphase representation as in Eq. (1.29), where we replace $X^{\downarrow N}_n(z)$ with $H^{\downarrow N}_n(z)$. Upsampling by $N = 2$ means replacing all z's by z^{-2}, as in Eq. (1.14).

So all together we have the following system or condition,

$$\left(H^{\downarrow N}_n(-z) \cdot H^{\downarrow N}_{N-1-n}(-z)\right) \downarrow 2 \uparrow 2 = delay$$

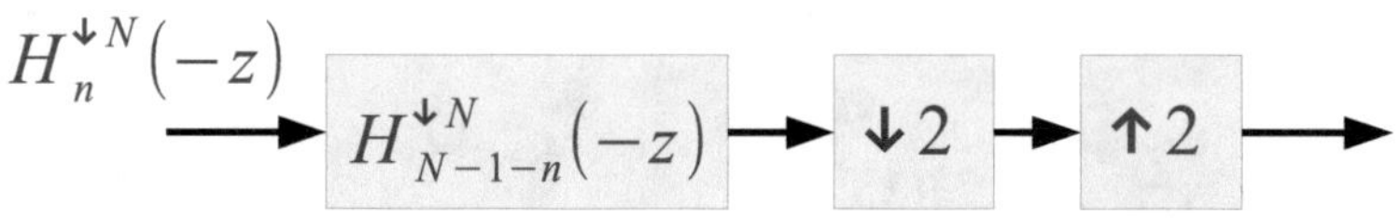

$$H_n^{\downarrow N}(-z)$$

Figure 1.33: A signal and filter with down- and up-sampling by a factor of 2.

Downsampling followed by upsampling with a factor of $N = 2$ is equivalent to multiplication with a delta train in the time domain. As we saw in Eq. (1.13) this causes the addition of alias components shifted by $2\pi/N = \pi$ in the frequency domain,

$$\frac{1}{2} \cdot H_n^{\downarrow N}(-z) \cdot H_{N-1-n}^{\downarrow N}(-z) + \frac{1}{2} \cdot H_n^{\downarrow N}(-z \cdot e^{j\pi}) \cdot H_{N-1-n}^{\downarrow N}(-z \cdot e^{j\pi}) = \pm delay \quad (1.73)$$

For the QMF, an assumption is that the prototype impulse response is symmetric around its centre. With the length of the impulse response of L we get $h(L - 1 - n) = h(n)$. Using (1.40) we obtain

$$H_{N-1-n}^{\downarrow N}(-z) = z^{-l+1} \cdot H_n^{\downarrow N}(-z^{-1}) \quad (1.74)$$

(with $l = L/N$). Using (1.74) we can reformulate (1.73) as an expression with the power spectrum (1.11 or 1.12), which is more easily to interpret,

$$|H_n^{\downarrow N}(e^{j\omega})|^2 + |H_n^{\downarrow N}(e^{j(\omega+\pi)})|^2 = 2 \quad (1.75)$$

Here we also got rid of the minus sign in front of the exponential function by using $-e^{j\omega} = e^{j(\omega-\pi)}$. Observe that this condition is now only for the unit circle in the complex z-plane, but using the so-called analytic continuation it can be shown that if it is true on the unit-circle, it is also true for the complex z-plane in general.

Our goal is to obtain bandpass filter with bandwidth π/N. Hence our prototype filter is a low pass, centred around zero, with its passband in the range of $-\pi/(2N) < \omega < \pi/(2N)$. The by a factor of N downsampled version $H_n^{\downarrow N}(e^{j\omega})$ has its passband accordingly (see (1.10)) in the range of $-\pi/2 < \omega < \pi/2$, which means it is a half band lowpass filter. Now we have a condition for the N polyphase components of our prototype filter, but we would like to have a condition for the full prototype filter. To get that, we can reformulate the polyphase components in terms of the full filter in the frequency domain, using our expression for downsampling (1.9).

$$H_n^{\downarrow N}(e^{j\Omega}) = \frac{1}{N} \sum_{k=0}^{N-1} e^{j\frac{2\pi}{N} \cdot k \cdot n} \cdot H(e^{j(-\frac{2\pi}{N} \cdot k + \frac{\Omega}{N})}) \quad (1.76)$$

Our prototype filters are real valued, hence the negative half of the spectrum is the conjugate complex version of the positive half, and we only need to consider the positive frequency axis $0 \le \omega < \pi$. The bandwidth of our filters should divide the positive frequency axis in N equal parts, hence we have a bandwidth of π/N and our prototype filter has its passband in the range of $\frac{-\pi}{2N}, \ldots, \frac{\pi}{2N}$ since its passband is centred around frequency 0. We now further assume that the filter has a transition width of 1 subband, such that it has a high attenuation at the next over subband, meaning

$$|H(e^{j\omega})| \approx 0 \quad \text{for} \quad 1.5\pi/N < |\omega| < \pi \tag{1.77}$$

In this case the sum in (1.76) has only one approximately non-zero element (for $k = 0$), and it becomes

$$H_n^{\downarrow N}(e^{j\Omega}) \approx \frac{1}{N} H(e^{j(\frac{\Omega}{N})}) \tag{1.78}$$

Now we can simplify (1.75) using (1.78), and use it as a condition on the positive frequency axis

$$|\frac{1}{N} \cdot H(e^{j\frac{\omega}{N}})|^2 + |\frac{1}{N} \cdot H(e^{j(\frac{\omega+\pi}{N})})|^2 = 2 \quad \text{for} \quad 0 \le \omega < \pi \tag{1.79}$$

or

$$|H(e^{j\frac{\omega}{N}})|^2 + |H(e^{j(\frac{\omega+\pi}{N})})|^2 = 2 \cdot N^2 \quad \text{for} \quad 0 \le \omega < \pi$$

This can also be seen as the sum of the magnitude squared frequency responses of 2 neighbouring subbands.

So all together we have these conditions or properties for the optimization process of a PQMF filter bank,

1. The analysis and synthesis prototype filters are identical (1.71),

$$H(z) = H'(z)$$

   and we assume near para-unitary polyphase matrices, hence

$$h(n) = h(LN - 1 - n),$$

   see (1.62).

2. The prototype filter needs high attenuation beyond the neighbouring subband (1.77),

$$|H(e^{j\omega})| \approx 0 \quad \text{for} \quad 1.5\pi/N < |\omega| < \pi$$

3. Two neighbouring subbands squared magnitude add to a constant (1.79),

$$|H(e^{j\frac{\omega}{N}})|^2 + |H(e^{j(\frac{\omega+\pi}{N})})|^2 \approx 2 \cdot N^2 \quad \text{for} \quad 0 \le \omega < \pi$$

Reconstruction Error of a PQMF Design

Now assume we found a PQMF design through optimization. Since the PQMF has no perfect reconstruction, it is useful to compute the reconstruction error of the obtained QMF design. How do we do this? If we multiply its analysis and synthesis polyphase matrix, perfect reconstruction would result in an identity matrix times a delay (the system delay minus the blocking delay). For our PQMF, it is this plus an error matrix, which contains error terms on the main diagonal (we could get the off-diagonal elements to zero by choosing the analysis and synthesis prototype identical).

Our Signal to Noise Ratio (SNR), where the noise is the reconstruction error, is

$$SNR = 10 \cdot \log_{10} \left(\frac{E(x(n)^2)}{E\left((x(n-n_d) - \hat{x}(n))^2\right)} \right).$$

This can be seen as the average signal power divided by the average error power. The error signal $x(n-n_d) - \hat{x}(n)$ in the z-domain is

$$\mathbf{X}(z) \cdot z^{-m_d} - \hat{\mathbf{X}}(z) = \mathbf{X}(z) \left(\mathbf{I} \cdot z^{-m_d} - \mathbf{H}(z) \cdot \mathbf{G}(z) \right).$$

Because of (1.71) we could make the off-diagonal elements of $\mathbf{H}(z) \cdot \mathbf{G}(z)$ equal to zero, and we can assume the diagonal elements as close to a delay. We can define a "residual" matrix as

$$\mathbf{R}(z) := \left(\mathbf{I} \cdot z^{-m_d} - \mathbf{H}(z) \cdot \mathbf{G}(z) \right) \tag{1.80}$$

$$= \text{diag} \left(R_0^{\downarrow N}(z), \ldots, R_{N-1}^{\downarrow N}(z) \right)$$

such that our error signal is

$$[X_0^{\downarrow N}(z) \cdot R_0^{\downarrow N}(z)), \ldots, X_{N-1}^{\downarrow N}(z) \cdot R_{N-1}^{\downarrow N}(z))]$$

To compute the average error power, we can use Parseval's theorem for the polyphase representation (1.42),

$$E(|x(n-n_d) - \hat{x}(n)|^2) = \frac{1}{2\pi} \int_{\omega=0}^{2\pi} \left| \sum_{n=0}^{N-1} e^{-j\omega n} \cdot X_n^{\downarrow N}(z) \cdot R_n^{\downarrow N}(e^{j\omega N}) \right|^2 d\omega$$

Now we assume that the spectrum of the polyphase components of our signal $X(e^{j\omega})$ are approximately constant and equal. This means we assume a more or less smooth spectrum for our signal. Then we have

$$P \approx X_{n'}^{\downarrow N}(e^{j\omega N})$$

with $|P|^2$ as the signal power of each polyphase element of the signal cdot P(which means the total signal power is $N \cdot P^2$), and hence

$$\sum_{n=0}^{\infty} |x(n-n_d) - \hat{x}(n)|^2 \approx |P|^2 \int_{\omega=0}^{2\pi} \left| \sum_{n=0}^{N-1} e^{-j\omega n} \cdot R_n^{\downarrow N}(e^{j\omega N})) \right|^2 d\omega$$

In non-polyphase representation this is

$$\sum_{n=0}^{\infty} |x(n-n_d) - \hat{x}(n)|^2 \approx |P|^2 \int_{\omega=0}^{2\pi} |R(e^{j\omega})|^2 d\omega$$

We can use Eq. (1.39) to obtain the non-polyphase representation of $\mathbf{R}(z)$ for its different phases, and from there its time-domain signal $r(n)$. Observe that $\mathbf{R}(z)$ only contains diagonal elements, hence $r(n)$ only contains the coefficients of the diagonal elements. Using the time-domain side of Parseval's theorem we get

$$\sum_{n=0}^{\infty} |x(n-n_d) - \hat{x}(n)|^2 \approx |P|^2 \sum_{n=0}^{\infty} |r(n)|^2$$

Hence the approximate SNR for the reconstruction error becomes

$$10 \log_{10} \left(\frac{N \cdot |P|^2}{|P|^2 \cdot \sum_{n=0}^{\infty} |r(n)|^2} \right) = 10 \log_{10} \left(\frac{N}{\sum_{n=0}^{\infty} |r(n)|^2} \right)$$

So all we need to evaluate our QMF design is to compute the energy of the diagonal elements of our residual matrix $R(z)$,

$$\sum_{n=0}^{\infty} |r(n)|^2$$

In conclusion, we simply compute the sum of squares of all coefficients of our residual matrix $R(z) = (\mathbf{I} \cdot z^{-m_d} - \mathbf{H}(z) \cdot \mathbf{G}(z))$ to obtain our estimate for the reconstruction error power of our QMF design.

Python Example for PQMF Filter Banks

The following is an example optimization function for a 4-band PQMF filter bank. It implements the above conditions 1–3 and combines them into a total error function. Condition 1 means we only need to design the first half of our prototype filter, and obtain the second half by mirroring. We store it in file "optimfuncQMF.py".

```python
from __future__ import print_function
def optimfuncQMF(x,N):
    """Optimization function for a PQMF Filterbank
    x: coefficients to optimize (first half of prototype h), N: Number of subbands
    err: resulting total error
    """

    import numpy as np
    import scipy as sp
    import scipy.signal as sig

    h = np.append(x,np.flipud(x));
    #H = sp.freqz(h,1,512,whole=True)
    f,H_im = sig.freqz(h)
    H=np.abs(H_im) #only keeping the real part
    posfreq = np.square(H[0:int(512/N)]);
    #Negative frequencies are symmetric around 0:
    negfreq = np.flipud(np.square(H[0:int(512/N)]))
```

```
        #Sum of magnitude squared frequency responses should be close to unity (or N)
        unitycond = np.sum(np.abs(posfreq+negfreq − 2*(N*N)*np.ones(int(512/N))))/512;
        #plt.plot(posfreq+negfreq);
        #High attenuation after the next subband:
        att = np.sum(np.abs(H[int(1.5*512/N):]))/512;
        #Total (weighted) error:
        err = unitycond + 100*att;
        return err
```

For the optimization we use the method "Sequential Least SQuares Programming"
(SLSQP) with a "good" starting point.

We add a main routine to the file to let the optimization run, store the result in the text
file "QMFcoeff.txt", and plot the resulting baseband impulse response and magnitude of
its frequency response (Figs. 1.34 and 1.35)

```
if __name__ == '__main__': #run the optimization
    from scipy.optimize import minimize
    import scipy as sp
    import matplotlib.pyplot as plt

    N=4 #Number of subbands
    #Start optimization with "good" starting point:
    x0 = 16*sp.ones(4*N)
    print("starting error=", optimfuncQMF(x0,N)) #test optim. function
    xmin = minimize(optimfuncQMF,x0, args=(N,), method='SLSQP')
    print("error after optim.=",xmin.fun)
    print("optimized coefficients=",xmin.x)
    #Store the found coefficients in a text file:
    sp.savetxt("QMFcoeff.txt", xmin.x)
    #we compute the resulting baseband prototype function:
    h = sp.concatenate((xmin.x,sp.flipud(xmin.x)))
    plt.plot(h)
    plt.xlabel('Sample')
    plt.ylabel('Value')
    plt.title('Baseband Impulse Response of the Optimized PQMF Filter Bank')
    #plt.xlim((0,31))
    plt.show()
    #The corresponding frequency response:
    w,H=sp.signal.freqz(h)
    plt.plot(w,20*sp.log10(abs(H)))
    plt.axis([0, 3.14, −100,20])
    plt.xlabel('Normalized Frequency')
    plt.ylabel('Magnitude (dB)')
    plt.title('Mag. Frequency Response of the PQMF Filter Bank')
    plt.show()
    #Checking the "unity condition":
    posfreq = sp.square(abs(H[0:int(512/N)]));
```

negfreq = sp.flipud(sp.square(**abs**(H[0:**int**(512/N)])))
plt.plot(posfreq+negfreq)
plt.xlabel('Frequency_(512_is_Nyquist)')
plt.ylabel('Magnitude')
plt.title('Unity_Condition,_Sum_of_Squared_Magnitude_of_2_Neigh._Subbands')
plt.show()

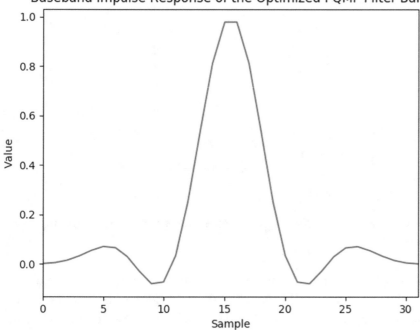

Figure 1.34: The PQMF baseband prototype impulse response resulting from our optimization.

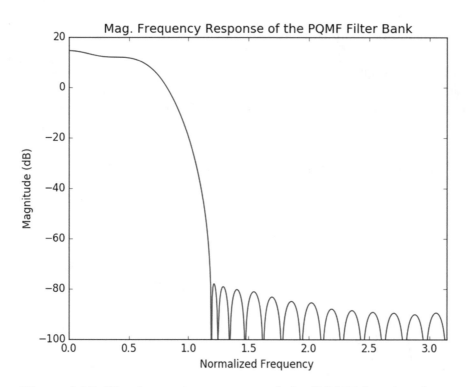

Figure 1.35: The frequency response of the PQMF baseband prototype resulting from our optimization.

Here in Fig. 1.35 we can inspect the pass band, from normalized frequency 0 to about 0.4, which has an attenuation of 0 to -3dB (compared to frequency 0) and next the transition band (which spans the neighbouring band) of width 0.78. The stop band starts at frequency $0.375\pi \approx 1.2$, and we can see that it has an attenuation of better than about -90 dB compared to the passband! This shows how much we fulfil the attenuation condition (1.77). We also check for the magnitude unity condition (1.79), with the main routine. The result is in Fig. 1.36. We see it is close to the desired value of $2N^2 = 32$, but we see deviations down to 31. This results in a not perfectly flat frequency response after reconstruction, and reminds us that we only have near perfect reconstruction.

Then we can implement the Pseudo Quadrature Mirror Filter bank, in file "PQMFB.py".

We need two functions. The first function constructs the analysis folding matrix, as in Eq. (1.59).

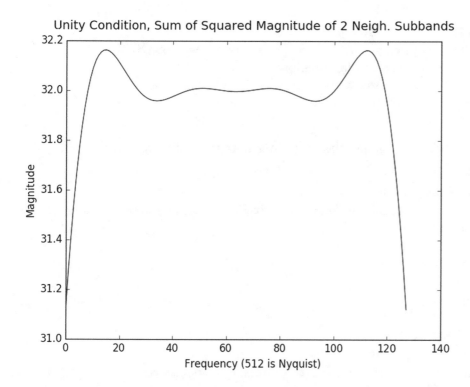

Figure 1.36: The resulting magnitude unity condition of our PQMF filter bank.

```
from __future__ import print_function
from polmatmult import polmatmult
from DCT4 import *
from x2polyphase import *

def ha2Fa3d(qmfwin,N):
    #usage: Fa=ha2Fa3d_fast(ha,N);
    #produces the analysis polyphase folding matrix Fa with all polyphase components
    #in 3D matrix representation
    #from a basband filter ha with
    #a cosine modulation
    #N: Blocklength
    #Matrix Fa according to
    #chapter about "Filter Banks", cosine modulated filter banks.

    overlap=int(len(qmfwin)/N)
    print("overlap=", overlap)
    Fa=np.zeros((N,N,overlap))
    for m in range(int(overlap/2)):
        Fa[:,:,2*m]+=np.fliplr(np.diag(np.flipud(
        -qmfwin[m*2*N:int(m*2*N+N/2)]*((-1)**m)),k=int(-N/2)))
        Fa[:,:,2*m]+=(np.diag(np.flipud(
        qmfwin[m*2*N+int(N/2):(m*2*N+N)]*((-1)**m)),k=int(N/2)))
```

```
        Fa[:,:,2*m+1]+=(np.diag(np.flipud(
        qmfwin[m*2*N+N:(m*2*N+int(1.5*N))]*((−1)**m)),k=−int(N/2)))
        Fa[:,:,2*m+1]+=np.fliplr(np.diag(np.flipud(
        qmfwin[m*2*N+int(1.5*N):(m*2*N+2*N)]*((−1)**m)),k=int(N/2)))
        #print −qmfwin[m*2*N:(m*2*N+N/2)]*((−1)**m)
    return Fa
```

The second function constructs the synthesis folding matrix, as in Eq. (1.60).

```
def hs2Fs3d(qmfwin,N):
    #usage: Fs=hs2Fs3d_fast(hs,N);
    #produces the synthesis polyphase folding matrix Fs with all polyphase components
    #in 3D matrix representation
    #from a basband filter ha with
    #a cosine modulation
    #N: Blocklength
    #Fast implementation

    #Fa=ha2Fa3d_fast(hs,N)
    #print "Fa.shape in hs2Fs : ", Fa.shape
    #Transpose first two dimensions to obtain synthesis folding matrix:
    #Fs=np.transpose(Fa, (1, 0, 2))
    overlap=int(len(qmfwin)/N)
    print("overlap=", overlap)
    Fs=np.zeros((N,N,overlap))
    for m in range(int(overlap/2)):
        Fs[:,:,2*m]+=np.fliplr(np.diag(np.flipud(
        qmfwin[m*2*N:int(m*2*N+N/2)]*((−1)**m)),k=int(N/2)))
        Fs[:,:,2*m]+=(np.diag((
        qmfwin[int(m*2*N+N/2):(m*2*N+N)]*((−1)**m)),k=int(N/2)))
        Fs[:,:,2*m+1]+=(np.diag((
        qmfwin[m*2*N+N:(m*2*N+int(1.5*N))]*((−1)**m)),k=int(−N/2)))
        Fs[:,:,2*m+1]+=np.fliplr(np.diag(np.flipud(
        −qmfwin[m*2*N+int(1.5*N):(m*2*N+2*N)]*((−1)**m)),k=int(−N/2)))
    #print "Fs.shape in hs2Fs : ", Fs.shape
    #avoid sign change after reconstruction:
    return −Fs
```

Now we can build the analysis filter bank function "PQMFBana",

```
def PQMFBana(x,N,fb):
    #Pseudo Quadrature Mirror analysis filter bank.
    #Arguments: x: input signal, e.g. audio signal, a 1−dim. array
    #N: number of subbands
    #fb: coefficients for the Quadrature filter bank.
    #returns y, consisting of blocks of subband in in a 2−d array of shape (N,# of blocks)

    Fa=ha2Fa3d(fb,N)
    print("Fa.shape=",Fa.shape)
```

```
y=x2polyphase(x,N)
print("y[:,:,0]=", y[:,:,0])
y=polmatmult(y,Fa)
y=DCT4(y)
#strip first dimension:
y=y[0,:,:]
return y
```

The synthesis filter bank function "PQMFBsyn" now becomes

```
from polyphase2x import *
def PQMFBsyn(y,fb):
    #Pseudo Quadrature Mirror synthesis filter bank.
    #Arguments: y: 2−d array of blocks of subbands, of shape (N, # of blokcs)
    #fb: prototype impulse response
    #returns xr, the reconstructed signal, a 1−d array.
    N, m=y.shape
    print("N=",N)
    Fs=hs2Fs3d(fb,N)
    #print np.transpose(Fs, axes=(2,0,1))
    #add first dimension to y for polmatmult:
    y=np.expand_dims(y,axis=0)
    xp=DCT4(y)
    xp=polmatmult(xp,Fs)
    xr=polyphase2x(xp)
    return xr
```

We again add a main routine to let the filter bank run, test, and evaluate the reconstruction error,

```
#Testing:
if __name__ == '__main__':
    import numpy as np
    import matplotlib.pyplot as plt

    #Number of subbands:
    N=4

    #D=Dmatrix(N)
    #Dinv=Dinvmatrix(N)
    #Filter bank coefficients, 1.5*N of sine window:
    #fb=np.sin(np.pi/(2*N)*(np.arange(int(1.5*N))+0.5))
    fb=np.loadtxt("QMFcoeff.txt")
    #compute the resulting baseband prototype function:
    fb = np.concatenate((fb,np.flipud(fb)))
    print("fb=", fb)
    plt.plot(fb)
    plt.title('The_PQMF_Protoype_Impulse_Response')
    plt.xlabel('Sample')
```

```
plt.ylabel('Value')
plt.xlim((0,31))
plt.show()
#input test signal, ramp:
x=np.arange(16*N)
plt.plot(x)
plt.title('Input_Signal')
plt.xlabel('Sample')
plt.ylabel('Value')
plt.show()
y=PQMFBana(x,N,fb)
plt.imshow(np.abs(y))
plt.title('PQMF_Subbands')
plt.xlabel('Block_No.')
plt.ylabel('Subband_No.')
plt.show()
xr=PQMFBsyn(y,fb)
plt.plot(xr)
plt.title('Reconstructed_Signal')
plt.xlabel('Sample')
plt.ylabel('Value')
plt.show()
#Input to the synthesis filter bank: unit pulse in lowest subband
#to see its impulse response:
y=np.zeros((N,2))
y[0,0]=1
xr=PQMFBsyn(y,fb)
plt.plot(xr)
plt.title('Impulse_Response_of_Modulated_Synthesis_Subband_0')
plt.xlabel('Sample')
plt.ylabel('Value')
plt.show()
#Check for reconstruction error:
#we can compute the residual matrix R from Fa and Fs:
Fa=ha2Fa3d(fb,N)
Fs=hs2Fs3d(fb,N)
R=polmatmult(Fa,Fs)
print( np.transpose(R, axes=(2,0,1)))
print("R[:,:,7]=", R[:,:,7])
R[:,:,7] = np.eye(N)-R[:,:,7]
#This is our residual matrix, from which we can compute the
#reconstruction error as the sum of all its squared diagonal
#elements,
re = 0;
for m in range(15):
```

We let it run with

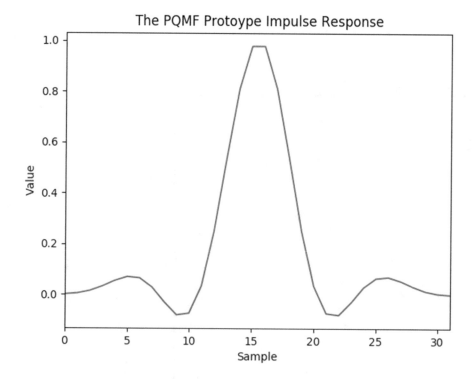

Figure 1.37: The PQMF prototype impulse response used for testing.

```
python PQMFB.py
```

which produces the plots in Figs. 1.37, 1.38, 1.39, 1.40, 1.41.

Observe Fig. 1.40. We can clearly see that our ramp signal is not perfectly reconstructed. Instead we see distortions on the order of perhaps 5%. This is the result of the PQMF having only a near perfect reconstruction. This means we need to evaluate our design not just for the quality of our obtained filters, but also if the reconstruction error fulfils our application requirements. Hence in the main routine we also evaluate the reconstruction error of our design. We use Eq. (1.80) to compute the residual matrix for the reconstruction error. Part of it are the analysis and synthesis polyphase matrices $\mathbf{H}(z)$ and $\mathbf{G}(z)$. To compute their product, we can also use $\mathbf{F_a}(z)$ and $\mathbf{F_s}(z)$, since

$$\mathbf{H}(z) \cdot \mathbf{G}(z) = \mathbf{F_a}(z) \cdot \mathbf{T} \cdot \mathbf{T}^{-1} \cdot \mathbf{F_s}(z) = \mathbf{F_a}(z) \cdot \mathbf{F_s}(z)$$

We compute the residual matrix $R(z)$ from (1.80),

```
Fa=ha2Fa3d(h,4);
Fs=hs2Fs3d(h,4);
R=polmatmult(Fa,Fs);
```

Since we have a near orthogonal filter bank, and a filter length of 32, we get a delay of 31 samples, or 28 samples (7 blocks of N=4 samples) without the blocking delay. Hence the matrix $\mathbf{I} \cdot z^{-m_d}$ in (1.80) has $m_d = 7$, such that we get

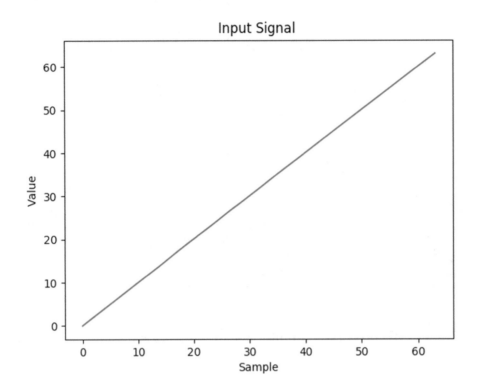

Figure 1.38: The ramp signal used for testing.

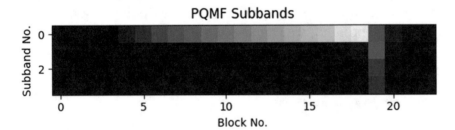

Figure 1.39: The PQMFB subbands resulting from the ramp signal as input. Observe that we have more subband blocks as the MDCT because of the longer filters.

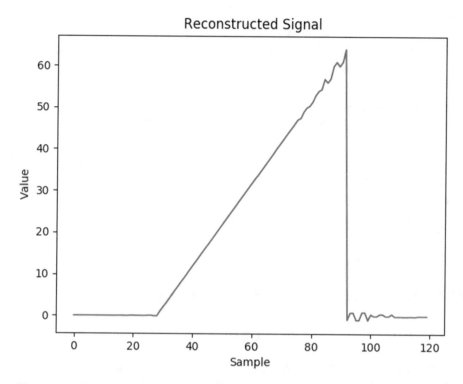

Figure 1.40: The reconstructed ramp function. Observe the distortions as a result of the PQMF only having a near perfect reconstruction.

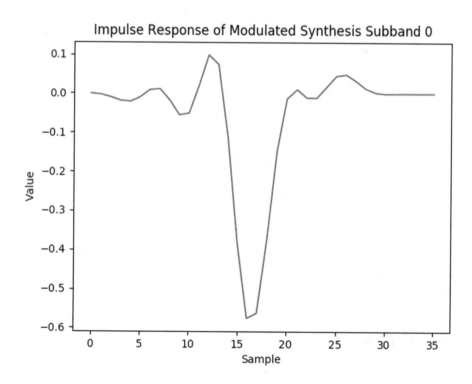

Figure 1.41: The impulse response of the synthesis filter of subband 0.

```
R[:,:,7] = eye(4)-R[:,:,7]
```

This is now our residual matrix, from which we can compute the reconstruction error as the sum of all its squared diagonal elements. The program outputs

```
Reconstruction error= 0.0069997943475
The resulting SNR for the reconstruction error= 27.5697471061 dB
```

which see that we get a reconstruction error of 0.00699, and an SNR of 27.57 dB, which corresponds to an error of $10^{(-27.57/20)}$ or 4.2%, which fits to our observation for the ramp signal. Observe that for audio coding purposes this would not be enough. There an SNR of clearly more than the maximum simultaneous masking threshold (see Sect. 4.4) is desirable.

1.15 Time-Varying and Switchable Filter Banks

1.15.1 With a Constant Number of Subbands

Using the polyphase representation we can now design time-varying filter banks, which means that the filter coefficients can change during processing of a signal in such a way that we still obtain perfect reconstruction. For that, the coefficients have to change in a certain way and with suitable timing between the analysis and synthesis filter bank.

To see how to do it, let's start with the simplest system, the processing of samples $x(m)$ (where m could be our block index) with just one time-varying coefficient h_m and a delay element z^{-1}, to produce an output. We see that multiplying $x(m)$ with h_m and then delaying it results in the same output as first delaying the signal $x(m)$ and then multiplying it by the delayed version of h_m,

$$x(m) \cdot h_m \cdot z^{-1} = x(m) \cdot z^{-1} \cdot h_{m-1} \tag{1.81}$$

(see also [31–33]). We can use this rule to obtain the inverse of a time varying system, like for a synthesis filter bank for perfect reconstruction. If we have the above system, we invert it with the inverse factor of the previous time step,

$$\left(x(m) \cdot h_m \cdot z^{-1}\right) \cdot \left(h_{m-1}^{-1}\right) = x(m-1) \tag{1.82}$$

This simple rule can now be applied to obtain a window switching during signal processing and with always perfect reconstruction.

1.15.2 Example: MDCT Window Switching

In Eq. (1.52) we saw that our MDCT analysis filter bank in our matrix description is

$$\mathbf{H}_{MDCT}(z) = \mathbf{F_a} \cdot \mathbf{D}(z) \cdot \mathbf{T}$$

and the synthesis is (1.54)

$$\mathbf{G}_{MDCT}(z) = \mathbf{T}^{-1} \cdot z^{-1} \cdot \mathbf{D}^{-1}(z) \cdot \mathbf{F_s}$$

Since we want to switch our window, the coefficients in the folding matrix $\mathbf{F}$ are time varying during the switching process. Our coefficients in $\mathbf{F}$ are $h_m(n)$, which define the window shape at block m, and hence we include this block index m as an argument in our folding matrix, $\mathbf{F}(m)$.

Hence, cascading analysis and synthesis filter banks we obtain the reconstructed signal

$$\hat{\mathbf{X}}(z) = \mathbf{X}(z) \cdot \mathbf{F_a}(m) \cdot \mathbf{D}(z) \cdot \mathbf{T} \cdot \mathbf{T}^{-1} \cdot z^{-1} \cdot \mathbf{D}^{-1}(z) \cdot \mathbf{F_s}(m) =$$

$$= \mathbf{X}(z) \cdot \mathbf{F_a}(m) \cdot z^{-1} \cdot \mathbf{F_s}(m)$$

Now Eq. (1.81) and (1.82) show us that we need

$$\mathbf{F_s}(m) = \mathbf{F_a}^{-1}(m-1)$$

for perfect reconstruction. Hence we get the time-varying analysis filter bank as

$$\mathbf{H}_{MDCT}(z, m) = \mathbf{F_a}(m) \cdot \mathbf{D}(z) \cdot \mathbf{T}(m)$$

and the time-varying synthesis filter bank for perfect reconstruction as

$$\mathbf{G}_{MDCT}(z, m) = \mathbf{T}^{-1}(m) \cdot z^{-1} \cdot \mathbf{D}^{-1}(z) \cdot \mathbf{F_a}^{-1}(m-1)$$

This shows us that we have to delay the windows switching of the synthesis window by 1 block relative to the analysis side. How does the resulting window shape or baseband impulse response look like during switching? So see that we need to note that only the later half of the baseband impulse response $h_m(n)$ of the analysis is affected by the delay elements after multiplication with the delay matrix $\mathbf{D}(z)$. This means at the switching block the first half of the baseband impulse response which is "seen" be the transform matrix $\mathbf{T}$ is already from the new shape, whereas the second half is from the old shape.

1.15.3 With a Changing Number of Subbands

If one (or more) of the windows for switching has a different number of subbands, we can still reformulate it into the previous case with a constant number of subbands. For that we take the maximum number of subbands as our block size. Blocks of a lower number of subbands can be increased in length by filling them up with zeros, until they reach this maximum block size. Assume $N - 1$ is this maximum block size, equal to the maximum number of subbands. N_2 is the smaller number of subbands. Then we obtain a filter bank with N_2 subbands and a block size of N_1 if we insert $N_1 - N_2$ zeros into the signal block on the analysis filter bank. On the synthesis side we then simply remove those zeros again. Hence we obtain the signal blocks $\mathbf{X}_2(z)$, with index "2" to indicate that this is for the case of N_2 subbands,

$$\mathbf{X}_2(z) = [\underbrace{0, \ldots, 0}_{(N_1-N_2)/2}, X_0(z), \ldots, \ldots, X_{N_2-1}(z), \underbrace{0, \ldots, 0}_{(N_1-N_2)/2}]$$

For the analysis side this means that the following polyphase matrix has no relevant entries in its outer rows since they are multiplied by zero. Hence these $N_1 - N_2$ rows can be set to zero. Since for critical sampling we want N_2 subbands, this means the final resulting polyphase matrix has only N_2 non-zero columns. For the synthesis side this means that the resulting polyphase matrix has $N_1 - N_2$ zero valued columns. In this way we now have a polyphase description for the mode with the lower number of subbands, but with the larger block size.

1.15.4 Example: MDCT and Low Delay Filter Bank Switching

On the analysis side, after multiplying this vector with the diamond shaped $\mathbf{F}$ matrix the zeros are shifted to centre of the result vector. Any following Zero-Delay matrices don't change the place of zeros any more, because of their bi-diagonal shape. Hence the corresponding entries in these matrices can be set to zero. The same is true for the transform matrix at the end of the processing chain in the analysis side.

On the synthesis side, the (inverse) transform matrix has the transposed form, with $N_1 - N_2$ zero valued columns in its centre. To indicate that these transform matrices are for the lower number of subbands N_2, we include the index "2". The analysis transform matrix is then $\mathbf{T}_{a2}$, and the synthesis transform $\mathbf{T}_{s2}$,

$$\mathbf{T}_{a2} = \begin{bmatrix} \mathbf{T}_u \\ \hline \mathbf{0} \\ \hline \mathbf{T}_d \end{bmatrix} \Big\} N_1 \underbrace{\qquad}_{N_2} \qquad \mathbf{T}_{s2} = \underbrace{[\ \mathbf{T}_l \mid \mathbf{0} \mid \mathbf{T}_r\ ]}_{N_1} \Big\} N_2$$

The folding matrices for the lower number of subbands have corresponding zeros in place in the centre of the matrix, with h_2 being the coefficients for the lower subband number N_2,

$$\mathbf{F}_{a2} = \begin{bmatrix} 0 & & 0 & 0 & & 0 \\ & h_2(2N_2-1) & & & h_2(N_2-1) & \\ h_2(1.5N_2) & & 0 & & & h_2(0.5N_2) \\ h_2(1.5N_2-1) & & & 0 & & -h_2(0.5N_2-1) \\ & h_2(N_2) & & & -h_2(0) & \\ 0 & & 0 & 0 & & 0 \end{bmatrix} \tag{1.83}$$

The synthesis folding matrix has the corresponding transpose form.

In this way we get only two steady states for the two numbers of subbands, and instead of the time index m we can simply use the indices "1" and "2", denoting the matrices for the two different subbands. For the larger subband number N_1 we get

$$\mathbf{Y}_1(z) = \mathbf{X}_1(z) \cdot \mathbf{F}_{a1} \cdot \mathbf{D}(z) \cdot \mathbf{T}_1$$

and for the synthesis

$$\hat{\mathbf{X}}_1(z) = \mathbf{Y}_1(z) \cdot \mathbf{T}_1^{-1} \cdot z^{-1} \cdot \mathbf{D}^{-1}(z) \cdot \mathbf{F}_{\mathbf{a}1}^{-1}$$

and correspondingly for the lower number of subbands N_2,

$$\mathbf{Y}_2(z) = \mathbf{X}_2(z) \cdot \mathbf{F}_{\mathbf{a}2} \cdot \mathbf{D}(z) \cdot \mathbf{T}_{\mathbf{a}2}$$

and for the synthesis

$$\hat{\mathbf{X}}_2(z) = \mathbf{Y}_2(z) \cdot \mathbf{T}_{\mathbf{s}2} \cdot z^{-1} \cdot \mathbf{D}^{-1}(z) \cdot \mathbf{F}_{\mathbf{a}2}^{-1}$$

The transition between the two states can be done in different ways. One way is to directly switch the analysis filter bank input from $\hat{\mathbf{X}}_1(z)$ to $\hat{\mathbf{X}}_2(z)$ and vice versa. But since the delay matrix $\mathbf{D}(z)$ delays only one half of a block, this would result in a mixed block with an intermediate number of subbands. This is usually not desired. To avoid this mixed block for the subbands, we can introduce a mixed block for the time signals. Since the transition happens at only one block each time, we use the time-domain description instead of the z-domain for it. Assume the transition from N_1 to N_2 subbands (switching down) happens at block m_{12} with first sample $x(n_{12})$, then we use the following transition block:

$$\mathbf{x}_{12}(m_{12}) = [x(n_{12}), \dots, \dots, x(n_{12} + (N_1 + N_2)/2 - 1), \underbrace{0, \dots, 0}_{(N_1 - N_2)/2}]$$

This block will still produce N_1 subbands, because of the delays in the delay matrix $\mathbf{D}(z)$. The next block in the sequence will have the shape of $\mathbf{X}_2(z)$ and produce N_2 subbands.

The analysis folding matrix for this transition has the form of $\mathbf{F}_{\mathbf{a}2}$, but with additional 1's to enable the full reconstruction of the block.

$$\mathbf{F}_{\mathbf{a}12} = \begin{bmatrix}
0 & & & & & 0 & 1 & & & & 0 \\
& & & & 0 & & & 1 & & & \\
& & h_2(2N_2-1) & & & & & h_2(N_2-1) & & & \\
& & & & & & & & & & \\
& h_2(1.5N_2) & & & & & & & h_2(0.5N_2) & & \\
& h_2(1.5N_2-1) & & & & & & & -h_2(0.5N_2-1) & & \\
& & & & & & & & & & \\
& & h_2(N_2) & & & & & -h_2(0) & & & \\
& & & 0 & & & & 0 & & & \\
& & & & & & & & & & \\
0 & & & & 0 & 0 & & & & & 0
\end{bmatrix} \quad (1.84)$$

The transform matrix for this transition folding matrix is still $\mathbf{T}_1$, which has the full $N_1 \times N_1$ shape.

The synthesis transition folding matrix has the corresponding transpose form. Observe that it has to be applied in the synthesis filter bank one block later because of the delay by one block of the Delay matrix and its inverse. For the same reason the transform matrix has to be switched one block later as in the analysis.

$$
\mathbf{F}_{s12} =
\begin{bmatrix}
0 & & & -h_2'(0.5N_2-1) & h_2'(0.5N_2) & & & & 0 \\
& {\cdot}^{\cdot}{}^{\cdot} & & & & & & {}^{\cdot}{}^{\cdot}{}_{\cdot} & \\
& -h_2'(0) & & & & & h'2(N_2-1) & & \\
& 0 & & & & & 0 & & \\
& {}_{\cdot}{}^{\cdot}{}^{\cdot} & & & & & & & {}^{\cdot}{}^{\cdot}{}_{\cdot} \\
0 & & & & & & & & 0 \\
1 & & & & & & & & 0 \\
& {}^{\cdot}{}_{\cdot}{}_{\cdot} & & & & & & & \\
& & 1 & & & & 0 & & \\
& & h_2'(N_2) & & & & h_2'(2N_2-1) & & \\
& & & {}^{\cdot}{}_{\cdot}{}_{\cdot} & & & & {\cdot}^{\cdot}{}^{\cdot} & \\
0 & & & h_2'(1.5N_2-1) & h'2(1.5N_2) & & & & 0
\end{bmatrix}
$$
$$(1.85)$$

Observe that the product of the analysis and synthesis transition folding matrices $\mathbf{F}_{a12} \cdot \mathbf{F}_{s12}$ is the identity matrix, except for the last $(N_1-N_2)/2$ entries, which are zero. This fits to the construction of our transition block.

Hence for the transition from the higher number of subbands N_1 to the lower number N_2 we obtain the following sequence of our polyphase products for analysis and synthesis,
$$\mathbf{x}_1(m_{12}) \cdot \mathbf{F}_{a12} \cdot \mathbf{D}(z) \cdot \mathbf{T}_1 \ \mathbf{T}_1^{-1} \mathbf{F}_{s12} \cdot \mathbf{D}(z)$$

Example

We would like to know the transition prototype impulse response of our analysis filter bank for switching down from N_1 to N_2 subbands. Assume the number of subbands to be $N_1 = 1024$ and $N_2 = 128$, as in the MPEG-AAC audio coder, and a sine window.

We see that the block reaching the transform consists of the left half, which is delayed by the Delay matrix $\mathbf{D}(z)$ and the right half, non-delayed. Hence Eq. (1.81) tells us that the left half contains the window coefficients of the previous block, for the larger N_1 subbands. The right half contains the coefficients of the analysis transition matrix Eq. (1.84). Comparing these with the generic folding matrix Eq. (1.59) we obtain the resulting prototype impulse response (Fig. 1.42). The beginning of this prototype impulse response comes from the right side of the transition folding matrix (which contains the first half of the sine window for N_2 subbands and the ones), the end of the prototype comes the from the analysis folding matrix for N_1 subbands (which contains the 2nd half of the sine window for N_1 subbands).

In Octave/Matlab:

```
h1=sin(pi/2048*(0:2047 +0.5));
h2=sin(pi/256*(0:255 +0.5));
```

```
hta=[h2(1:128),ones(1,(1024-128)/2), h1(1025:2048)];
plot(hta)
```

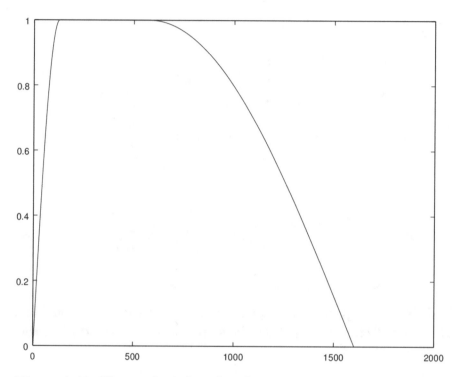

Figure 1.42: The analysis baseband prototype impulse response for the transition from 1024 to 128 bands. Remember that for the analysis the window function is the time-reversed version.

For the synthesis side be can imaging applying a pulse at the switching instance at the lowest subband. It then appears undelayed by the Delay matrix at the upper half of our synthesis transition matrix Eq. (1.85). After that it appears at the lower half of the matrix. Assume that the current folding matrix is for N_1 subbands, and the next one is the synthesis transition folding matrix $\mathbf{F_{s12}}$ Eq. (1.85). Then the resulting prototype for the switching consists of the first half of the prototype for N_1 subbands, followed by the ones and the second half of the prototype for N_2 subbands (the lower half of our transition folding matrix) (Fig. 1.43). Instead of reading out the impulse response we again just read out the prototype by comparing to the generic folding matrix for the synthesis Eq. (1.60).

In Matlab/Octave:

```
h1=sin(pi/2048*(0:2047 +0.5));
h2=sin(pi/256*(0:255 +0.5));
hts=[h1(1:1024),ones(1,(1024-128)/2), h2(129:256)];
plot(hts)
```

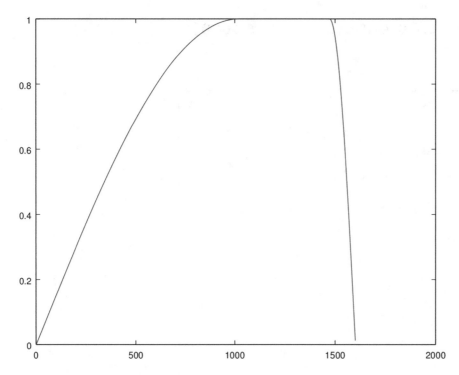

Figure 1.43: The synthesis baseband prototype impulse response for the transition from 1024 to 128 bands. Remember that for the synthesis this is identical to the window function.

switchdownirsynt.png

The analysis transition folding matrix for switching up, from N_2 to N_1 subbands, has a shape similar to Eq. (1.84),

$$
\mathbf{F_{a21}} =
\begin{bmatrix}
0 & & & 0 & 0 & & & 0 \\
& \ddots & 0 & & & 0 & \ddots & \\
& h_2(2N_2-1) & & & & & h_2(N_2-1) & \\
& \ddots & & & & & \ddots & \\
h_2(1.5N_2) & & & & & & & h_2(0.5N_2) \\
h_2(1.5N_2-1) & & & & & & & -h_2(0.5N_2-1) \\
& \ddots & & & & & \ddots & \\
& h_2(N_2) & & & & & -h_2(0) & \\
& & 1 & & & 0 & & \\
& & \ddots & & \ddots & & & \\
0 & & & 1 & 0 & & & 0
\end{bmatrix}
\tag{1.86}
$$

For the synthesis the transition folding matrix again has the transpose form,

$$
\mathbf{F}_{\mathbf{s}21} =
\begin{bmatrix}
0 & & & -h_2'(0.5N_2-1) & h_2'(0.5N_2) & & & & 0 \\
 & & \cdot{}^{\cdot}{}^{\cdot} & & & & & \ddots & \\
 & -h_2'(0) & & & & & h'2(N_2-1) & & \\
 & 0 & & & & & & 1 & \\
 & \cdot{}^{\cdot}{}^{\cdot} & & & & & & & \ddots \\
0 & & & & & & & & 1 \\
0 & & & & & & & & 0 \\
 \ddots & & & & & & & & \\
 & 0 & & & & & & 0 & \\
 & h_2'(N_2) & & & & & h_2'(2N_2-1) & & \\
 & \ddots & & & & & & \cdot{}^{\cdot}{}^{\cdot} & \\
0 & & & h_2'(1.5N_2-1) & h'2(1.5N_2) & & & & 0
\end{bmatrix}
$$

$$(1.87)$$

For transitioning back from the lower number of subbands N_2 to the higher number N_1, at block number m_{21} with first sample n_{21}, we use the transition block

$$
\mathbf{x}_{21}(m_{21}) = [\,\underbrace{0,\ldots,0}_{(N_1-N_2)/2}, x(n_{21}),\ldots,\ldots, x(n_{21}+(N_1+N_2)/2-1)]
$$

This block will still produce N_2 subbands, but the following block of the shape of $\mathbf{X}_2(z)$ will produce N_1 subbands, with transform matrix $\mathbf{T}_{\mathbf{a}2}$. The synthesis filter bank will reconstruct this block (as can be seen from the perfect reconstruction property and the system delay), but again with one block delay.

References

1. M. Kahrs, K. Brandenburg (eds.), *Applications of Digital Signal Processing to Audio and Acoustics* (Springer, New York, 1998)

2. J. Benesty, M.M. Sondhi, Y. Huang (eds.), *Handbook of Speech Processing* (Springer, New York, 2008)

3. J. Watkinson, *The MPEG Handbook* (Elsevier, Oxford, 2001)

4. P.P. Vaidyanathan, *Multirate Systems and Filter Banks* (Prentice-Hall, Upper Saddle River, NJ, USA, 1993)

5. N. Fliege, *Multirate Digital Signal Processing* (Wiley, New York, USA, 1994)

6. A.V. Oppenheim, R.W. Schafer, *Discrete-Time Signal Processing.* Signal Processing Series (Prentice-Hall, Upper Saddle River, NJ, USA, 2014)

7. R.N. Bracewell, *The Fourier Transform and Its Applications* (McGraw-Hill, New York, 1986)

8. A. Papoulis, *Signal Analysis* (McGraw-Hill, New York, 1984)

9. K.R. Rao, P. Yip, *The Transform and Data Compression Handbook* (CRC Press, Boca Raton, FL, USA, 2000)

10. A. Mertins, D.A. Mertins, *Signal Analysis: Wavelets, Filter Banks, Time-Frequency Transforms and Applications* (Wiley, New York, USA, 1999)

11. J.S. Lim, A.V. Oppenheim (eds.), *Advanced Topics in Signal Processing* (Prentice-Hall, Upper Saddle River, NJ, USA, 1987)

12. A. Spanias, T. Painter, V. Atti, *Audio Signal Processing and Coding* (Wiley, New York, USA, 2005)

13. A. Papandreou-Suppappola (ed.), *Applications in Time-Frequency Signal Processing* (CRC Press, Boca Raton, FL, USA, 2003)

14. M.M. Goodwin (ed.), *Adaptive Signal Models: Theory, Algorithms and Audio Applications* (Kluwer Academic Publishers, Boston, 1998)

15. M. Bellanger, *Digital Processing of Signals* (Wiley, New York, USA, 1984)

16. J.P. Princen, A.B. Bradley, Analysis/synthesis filter bank design based on time domain alias cancellation. IEEE Trans. Acoust. Speech Signal Process. 1153–1161 (1986)

17. G. Strang, T.Q. Nguyen, *Wavelets and Filter Banks* (Wellesley-Cambridge Press, Wellesley, 1996)

18. H.S. Malvar, *Signal Processing with Lapped Transforms* (Artech House, Norwood, MA, USA, 1992)

19. I. Bronshtein, K. Semendyayev, G. Musiol, H. Mühlig, *Handbook of Mathematics* (Springer, Berlin, Heidelberg, 2007)

20. G. Schuller, M.J.T. Smith, New framework for modulated perfect reconstruction filter banks. IEEE Trans. Signal Process. **44**, 1941–1954 (1996)

21. T.Q. Nguyen, R.D. Koilpillai, The theory and design of arbitrary length cosine-modulated filter banks and wavelets, satisfying perfect reconstruction. IEEE Trans. Signal Process. **44**, 473–483 (1996)

22. M. Schnell, R. Geiger, M. Schmidt, M. Jander, M. Multrus, G. Schuller, J. Herre, Mpeg-4 enhanced low delay AAC - low bitrate high quality communication, in *122nd AES Convention, Vienna, Austria*, May 2007

23. M. Schnell, R. Geiger, M. Schmidt, M. Multrus, M. Mellar, J. Herre, G. Schuller, Low delay filterbanks for enhanced low delay audio coding, in *IEEE Workshop on Applications of Signal Processing to Audio and Acoustics, New Paltz, NY*, 2007

24. G. Schuller, T. Karp, Modulated filter banks with arbitrary system delay: Efficient implementations and the time-varying case. IEEE Trans. Signal Process. **48**, 737–748 (2000)

25. G. Schuller, *Zeitvariante Filterbaenke mit niedriger Systemverzoegerung und perfekter Rekonstruktion* (VDI Verlag GmbH, Duesseldorf, 1999)

26. J. Breebart, C. Faller, *Spatial Audio Processing: MPEG Surround and Other Applications* (Wiley, New York, USA, 2007)

27. E. Larsen, R.M. Aarts, *Audio Bandwidth Extension* (Wiley, New York, USA, 2004)

28. P. Chu, Quadrature mirror filter design for an arbitrary number of equal bandwidth channels. IEEE Trans. Acoust. Speech Signal Process. 203–218 (1985)

29. T. Nguyen, Near-perfect-reconstruction pseudo-QMF banks. IEEE Trans. Signal Process. **43**, 65–76 (1994)

30. J.O. Smith III, J.S. Abel, *Spectral Audio Signal Processing* (W3K Publishing, Standford, 2011)

31. S. Phoong, P. Vaidyanathan, A polyphase approach to time-varying filter banks, in *Proc. of the IEEE International Conference on Acoustics, Speech and Signal Processing (ICASSP)*, vol. 3, pp. 1554–1557 (Atlanta, May 1996)

32. S. Phoong, P. Vaidyanathan, Time-varying filters and filter banks: Some basic principles. IEEE Trans. Signal Process. **44**, (1996)

33. G. Schuller, Time-varying filter banks with variable system delay, in *Proc. of the IEEE International Conference on Acoustics, Speech and Signal Processing (ICASSP)* (Munich, Germany), April 21–24, 1997

2 Quantization

Quantization is an important step towards a compact binary representation, and it is part of the irrelevance reduction, which means we lose information. Our goal is that this information loss is inaudible to the listener of the decoded audio signal, and we hope to achieve this goal by using a psycho-acoustic model which computes the masking thresholds for our signal. Since this threshold will tell us where the sensitivity threshold of the ear for each frequency in the presence of our signal is, we can choose the quantization step size such that the corresponding quantization error stays below this masking threshold. In this way, the quantization "noise" will hopefully be inaudible by the listener [1–4].

A **uniform "Mid-Tread" quantization** of a value or sample y with quantization step size Δ is defined as (in Python)

$$y_q = round(y/\Delta), \tag{2.1}$$

where "round" is the usual rounding to the nearest integer, and yq is the quantization index, meaning it is a (possibly signed) integer. This index will then be represented by binary representations using entropy coding in the encoder. This type of quantizer is used, for instance, in the MPEG-1/2 Layer I and II, ATRAC, and AC-3 audio coders.

In the decoder, this index is obtained by the entropy decoder from the binary representation, and this index yq is then converted into the de-quantized value using

$$y_{deq} = y_q \cdot \Delta \tag{2.2}$$

The quantization error is defined as

$$e = y_{deq} - y$$

and this quantization error is in the range of

$$-\Delta/2 \le e \le \Delta/2$$

To determine if our quantization error is below our masking threshold, we need to compute the average or expected quantization error power $E(e^2)$. For that it is helpful to assume that our signal y is much bigger than our quantization step size Δ (but observe that especially at low bit rates this will not be realistic). If this assumption is true, we can assume that each quantization error is equally likely. In that case, the probability density function of our quantization error $p(e)$ is

$$p(e) = 1/\Delta$$

© Springer Nature Switzerland AG 2020
G. Schuller, *Filter Banks and Audio Coding*,
https://doi.org/10.1007/978-3-030-51249-1_2

for $-\Delta/2 \le e \le \Delta/2$, and 0 outside this range. Hence the average quantization error power is computed as

$$E(e^2) = \int_{-\Delta/2}^{\Delta/2} p(e) \cdot e^2 de = \int_{-\Delta/2}^{\Delta/2} e^2/\Delta \cdot de$$

$$= 1/\Delta \cdot (((\Delta/2)^3)/3 - ((-\Delta/2))^3) = \frac{\Delta^2}{12} \quad (2.3)$$

In this way, uniform quantization gives us a precise control of the quantization distortions or "noise".

Now assume that instead of knowing the quantization step size Δ, we know the desired number of bits N for a fixed size binary number for the quantization index y_q, which means we can get up to 2^N different quantization indices or "quantization intervals". Our signal $x(n)$ is assumed to be between $-A/2$ and $+A/2$, and hence has a range of $A/2 - (-A/2) = A$. If we assume a fixed or uniform quantization step size, it then results to

$$\Delta = \frac{A}{2^N} \quad (2.4)$$

Let us further assume that our signal is uniformly distributed on its range (a not very realistic but mathematically convenient assumption), then the signal power can be computed like the quantization error power (2.3),

$$E(x^2) = \frac{A^2}{12} \quad (2.5)$$

Combining (2.3), (2.4), and (2.5), this then leads to a Signal to Error or Signal to Noise Ratio of

$$\frac{E(x^2)}{E(e^2)} = \frac{A^2/12}{\Delta^2/12} = \frac{A^2}{(A/2^N)^2} = \frac{1}{(1/2^N)^2} = 2^{2N}$$

or in dB,

$$SNR = 10 \cdot \log_{10}(2^{2N}) = N \cdot 2\log_{10}(2) \approx N \cdot 6.02\,\text{dB} \quad (2.6)$$

Equation (2.6) now says that each additional bit for the quantization index gives us an additional 6 dB of SNR, or a doubling in the power domain, which in practice often is clearly audible.

A modified quantization scheme, which contributes to the quantization noise shaping, is **non-uniform quantization**. In MPEG-1/2 Layer 3 and MPEG-2/4 AAC it is defined as (in Python) [1, 5, 6]

$$y_q = round(y^{0.75}/\Delta)$$

This leads to larger quantization step size for larger values, which are usually less sensitive to it, and smaller quantization step sizes for smaller values, which are usually more sensitive to it.

The de-quantization is

$$y_{deq} = (y_q \cdot \Delta)^{1/0.75}$$

106

For the de-quantization, in any case the decoder needs that quantization step size *delta*, which the encoder needs to send to the decoder as side information. To minimize this side information, the quantization step size is chosen to be the same for defined groups of subbands, called "scalefactor-bands", for instance, in the MPEG-1/2 Layer 3, and the MPEG-2/4 AAC coders. The scalefactor then takes the role of the quantization step size *delta* [5]. Often, the scalefactor-bands have a width of about 1/2 or 1/3 Bark band (see also Sect. 4.2).

These scalefactors also need to be quantized, and often a **logarithmic quantizer** is used, as in MPEG-AAC [5]. In the encoder we get the quantized scalefactors as

$$sf_q = round(\log_2(sf) \cdot 4)$$

and in the decoder,

$$sf_{deq} = 2^{(sf_q/4)}$$

References

1. M. Bosi, R.E. Goldberg, *Introduction to Digital Audio Coding an Standards* (Kluwer Academic Publishers, Dordrecht, 2003)

2. N.S. Jayant, P. Noll, *Digital Coding of Waveforms: Principles and Applications to Speech and Video* (Prentice Hall, Englewood, 1990)

3. J. Watkinson, *The MPEG Handbook* (Elsevier, Amsterdam, 2001)

4. Y. You, *Audio Coding* (Springer, Berlin, 2010)

5. MPEG, *Information Technology: Generic Coding of Moving Pictures and Associated Audio Information—Part 7: Advanced Audio Coding (AAC)* (ISO/IEC, 1997)

6. F.C. Pereira, T. Ebrahimi (eds.), *The MPEG-4 Book* (Prentice Hall PTR, Englewood, 2002)

3 Predictive Coding

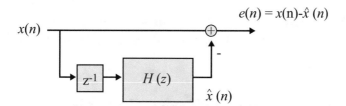

Figure 3.1: A block diagram of a predictive encoder. Observe that the predictor has only access to past samples through the delay element z^{-1}.

3.1 Introduction

Predictive audio coding uses the fact that neighbouring audio samples are usually quite similar, and often follow a more or less smooth straight or curved line. Hence it should be possible to predict the coming sample from previous samples, for instance, using extrapolation, which we call "prediction". We call the algorithm which computes this predicted value a "predictor", and the difference between the predicted value and the true coming value the "prediction error".

If we have a "good" prediction, then this prediction error will be small on average for our audio signal. Hence we can obtain a reduced bit rate by encoding and transmitting this prediction error instead of the original audio sample. Let $\hat{x}(n)$ be the predicted value, as output of the predictor at time, and $x(n)$ the true sample. Then the encoder computes and transmits the prediction error (Fig. 3.1)

$$e(n) := x(n) - \hat{x}(n). \tag{3.1}$$

This works because in the decoder we can obtain the reconstructed original audio sample $x_{rec}(n)$ by also computing the predicted value, based on the past reconstructed samples, and by adding the received prediction error to the current predicted value (Fig. 3.2),

$$x_{rec}(n) = e(n) + \hat{x}(n) \tag{3.2}$$

© Springer Nature Switzerland AG 2020
G. Schuller, *Filter Banks and Audio Coding,*
https://doi.org/10.1007/978-3-030-51249-1_3

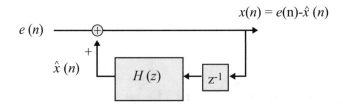

Figure 3.2: A block diagram of a predictive decoder. Observe that the predictor here also has access to the past samples through the reconstruction.

For this to work we just have to make sure that the predictors in the encoder and decoder produce exactly the same predicted values $\hat{x}(n)$, otherwise encoder and decoder would diverge.

3.2 The Mean Squared Error Solution

How do we obtain such a predictor? We make the model assumption that our audio signal behaves like the output of a generating system which consists of a random noise generator followed by an IIR filter, with an unknown number of poles in the z-domain of its transfer function. This is more or less realistic, for instance, for flutes or trumpets, whose input is a more or less noise-like signal at the mouthpiece. Their tube structure and the reflections back and forth inside this structure can be modeled by an IIR filter. The IIR filters feedback path models those reflections. This model also fits to unvoiced speech, where the vocal tract can be seen as a tube-like structure with reflections with noise-like input or excitation. For voiced speech the input or excitation signal would be periodic pulses instead of noise.

Now if we assume this signal model, we can cancel the poles of the transfer function of its generating system by a filter which has zeros at their positions in the z-domain of the transfer function. This is also convenient because it means the prediction error filter (the filter producing the prediction error) is simply a finite impulse response (FIR) filter. What should be left after such an FIR filter is simply the excitation noise (or periodic pulses in the case of voiced speech). Ideally the resulting noise would be unpredictable, and we would be done with the prediction.

These considerations show us how the basic structure of a predictor could look like. The output of our FIR prediction error filter is the prediction error, and the prediction error is the difference between the true coming sample and the predicted sample. Hence the predictor must also be an FIR filter. The problem left to solve is that we do not know the zero positions of the transfer function of the prediction error filter. But we do know how to recognize the perfect predictor: it minimizes the prediction error $e(n)$ in some sense. To be able to apply known mathematical tools, we choose the mean square prediction error (MSE) as the criterion to minimize with our predictor. The mean squared error (MSE) is defined as

$$E(e^2(n)),$$

where $E(.)$ is the "Expectation", for our purposes the average over all samples. We assume that our FIR predictor filter $h(n)$ has L coefficients (L has to be chosen later). Then the output of our predictor is computed from the past L audio samples using the following convolution of the past signal samples (before the present sample $x(n)$) with the predictor filter impulse response,

$$\hat{x}(n) = \sum_{m=0}^{L-1} x(n-1-m) \cdot h(m) \tag{3.3}$$

A well-known mathematical approach to obtain the minimum of a mean squared error in a matrix framework is the so-called Moore–Penrose Pseudo-Inverse [1]. To be able to apply it we can reformulate our convolution Eq. (3.3) as a matrix multiplication. The following is a formulation for the predictor output starting at sample $n = L$ to avoid samples with negative indices in the matrix,

$$\underbrace{\begin{bmatrix} x(L-1) & x(L-2) & \cdots & x(0) \\ x(L) & x(L-1) & \cdots & x(1) \\ x(L+1) & x(L) & \cdots & x(2) \\ & \vdots & & \vdots \\ x(B-1) & x(B-2) & \cdots & x(B-L) \end{bmatrix}}_{=:\mathbf{A}} \cdot \underbrace{\begin{bmatrix} h(0) \\ h(1) \\ \vdots \\ h(L-1) \end{bmatrix}}_{=:\mathbf{h}} = \underbrace{\begin{bmatrix} \hat{x}(L) \\ \hat{x}(L+1) \\ \hat{x}(L+2) \\ \vdots \\ \hat{x}(B) \end{bmatrix}}_{=:\hat{\mathbf{x}}}, \tag{3.4}$$

where B is the "block length", the number of samples chosen to compute the predictor coefficients $\mathbf{h}$. Observe that the matrix $\mathbf{A}$ in Eq. (3.4) has rows which are right shifted versions, by 1 sample, of the row above. This kind of matrix is called a "Toeplitz matrix" [2]. We now want these predicted values in $\hat{\mathbf{x}}$ to approximate the true values in the vector

$$\mathbf{x} := \begin{bmatrix} x(L) \\ x(L+1) \\ x(L+2) \\ \vdots \\ x(B) \end{bmatrix}$$

We can formulate this approximation in matrix terms as

$$\mathbf{A} \cdot \mathbf{h} \to \mathbf{x}$$

Observe that $\mathbf{A}$ is a tall matrix, meaning it has many rows and a few columns. Hence it is over determined, meaning it has fewer variables in $\mathbf{h}$ than it has rows, such that we can only get an approximation and not an equality. The Moore–Penrose Pseudo-Inverse now gives as the solution for the case that we want to minimize the mean squared error between $\hat{\mathbf{x}}$ and $\mathbf{x}$ using the following steps. First we multiply both sides with the transpose $\mathbf{A^T}$. This now makes the number of rows equal to the number of variables in $\mathbf{h}$, such that we can obtain the quality here, and hence we can use the equal sign "=",

$$\mathbf{A^T} \cdot \mathbf{A} \cdot \mathbf{h} = \mathbf{A^T} \cdot \mathbf{x} \tag{3.5}$$

Provided that the resulting matrix on the left-hand side is invertible, we can bring it to the right-hand side to obtain the solution for $\mathbf{h}$,

$$\mathbf{h} = (\mathbf{A^T} \cdot \mathbf{A})^{-1} \cdot \mathbf{A^T} \cdot \mathbf{x} \tag{3.6}$$

This is now our desired solution for the minimum mean squared error case.

If we let our block length go to infinity, $\mathbf{B} \to \infty$, then this becomes the well-known "Wiener–Hopf" equation,

$$\lim_{B \to \infty} \mathbf{A^T} \cdot \mathbf{A} = \underbrace{\begin{bmatrix} r_{xx}(0) & x(1) & \cdots & r_{xx}(L-1) \\ r_{xx}(1) & r_{xx}(0) & \cdots & r_{xx}(L-2) \\ r_{xx}(2) & r_{xx}(1) & \cdots & r_{xx}(L-3) \\ & \vdots & & \vdots \\ r_{xx}(L-1) & r_{xx}(L-2) & \cdots & r_{xx}(0) \end{bmatrix}}_{=:\mathbf{R_{xx}}}$$

with

$$r_{xx}(n) := \sum_{m=-\infty}^{\infty} x(m) \cdot x(m-n)$$

Observe that this is also a Toeplitz matrix. Further we get

$$\lim_{B \to \infty} \mathbf{A^T} \cdot \mathbf{x} = \underbrace{\begin{bmatrix} r_{xx}(1) \\ r_{xx}(2) \\ r_{xx}(3) \\ \vdots \\ r_{xx}(L) \end{bmatrix}}_{=:\mathbf{r_{xx}}}$$

Hence for $B \to \infty$ Eq. (3.5) turns into

$$\mathbf{R_{xx}} \cdot \mathbf{h} = \mathbf{r_{xx}}$$

and we get the solution

$$\mathbf{h} = \mathbf{R_{xx}}^{-1} \cdot \mathbf{r_{xx}}$$

This matrix form is also called the "Yule–Walker equation", and the general statistical formulation is the Wiener–Hopf equation [3]. This formulation has the advantage that it is a simple statistical formulation, with only correlations as input. The formulation with the pseudo-inverse has the advantage that it can be used with finite length data, which is more realistic.

Our prediction error, using Eq. (3.3), is now

$$e(n) = x(n) - \hat{x}(n) = x(n) - \sum_{m=1}^{L-1} x(n-m) \cdot h(m-1) = x(n) * [1, -\mathbf{h}^T]$$

with the predictor error filter coefficients $h_{perr}(n) := [1, -\mathbf{h}^T]$. In the z-domain this is

$$E(z) = X(z) \cdot \left(1 - z^{-1}H(z)\right) = X(z) \cdot H_{perr}(z) \tag{3.7}$$

The decoder computes the reconstructed signal (3.2)

$$x_{rec}(n) = e(n) + \hat{x}(n) = e(n) + \sum_{m=1}^{L-1} x_{rec}(n-m) \cdot h(m-1) \tag{3.8}$$

We can see that the reconstruction filter in the decoder is an IIR filter, since its input contains past reconstructed samples. IIR filters are easily analysed in the z-domain,

$$X_{rec}(z) = E(z) + z^{-1}X_{rec}(z) \cdot H(z)$$

or

$$X_{rec}(z) = E(z) \cdot \frac{1}{1 - z^{-1}H(z)} = E(z) \cdot \frac{1}{Hperr(z)}$$

We see that the reconstruction filter in the z-domain is exactly the inverse of the prediction error filter in the encoder, and hence the decoder exactly cancels the prediction error filter of the encoder for the signal reconstruction, as intended.

3.3 Online Adaptation

When we look at audio signal waveforms, we see that their characteristics and hence their statistics is changing, they are not stationary. Hence we can expect a prediction improvement if we divide an audio or speech signal into small pieces for the computation of the prediction coefficients.

3.3.1 LPC Coder

Especially for predicting and coding speech signals, the prediction coefficients are computed for blocks of about 20 ms duration, because within these blocks the signal statistics (the correlation coefficients) are assumed to be sufficiently stable. For each of these blocks, both the prediction error and the used prediction coefficients are then transmitted to the decoder for the reconstruction. This is called "Linear Predictive Coding" or short "LPC" [4–7].

Python Example

If we use $32\,\mathrm{kHz}$ sampling rate, $20\,\mathrm{ms}$ corresponds to 640 samples. The following example shows a for loop which runs over these blocks, and for each of those constructs the "A" matrices, computes the prediction coefficients $\mathbf{h}$ according to Eq. (3.6), the prediction error filter h_{perr}, and with it the prediction error $e(n)$ for each block. The linear filter function "lfilter" uses the prediction error filter h_{perr} in the numerator to compute the prediction error. The array "state" of "lfilter" contains the filter memory, and is necessary to make the predictor work across block boundaries.

```
L=10 #predictor lenth
len0 = np.max(np.size(x))
e = np.zeros(np.size(x)) #prediction error variable initialization
blocks = np.int(np.floor(len0/640)) #total number of blocks
state = np.zeros(L) #Memory state of prediction filter
#Building our Matrix A from blocks of length 640 samples and process:
h=np.zeros((blocks,L)) #initialize pred. coeff memory
```

```
for m in range(0,blocks):
    A = np.zeros((640−L,L)) #trick: up to 630 to avoid zeros in the matrix
    for n in range(0,640−L):
        A[n,:] = np.flipud(x[m*640+n+np.arange(L)])

    #Construct our desired target signal d, one sample into the future:
    d=x[m*640+np.arange(L,640)];
    #Compute the prediction filter:
    h[m,:] = np.dot(np.dot(np.linalg.inv(np.dot(A.transpose(),A)), A.transpose()), d)
    hperr = np.hstack([1, −h[m,:]])
    e[m*640+np.arange(0,640)], state = sp.lfilter(hperr,[1],x[m*640+np.arange(0,640)], zi=
        state)
```

The example for the decoder, implementing Eq. (3.8) or Eq. (3.7) implements the reconstruction filter again with the function "lfilter", but with h_{perr} in the denominator. Observe that the decoder uses both the prediction coefficients in array "hperr" and the prediction error in array "e" for the reconstruction.

```
for m in range(0,blocks):
    hperr = np.hstack([1, −h[m,:]])
    #predictive reconstruction filter: hperr from numerator to denominator:
    xrek[m*640+np.arange(0,640)] , state = sp.lfilter([1], hperr,e[m*640+np.arange(0,640)],
        zi=state)
```

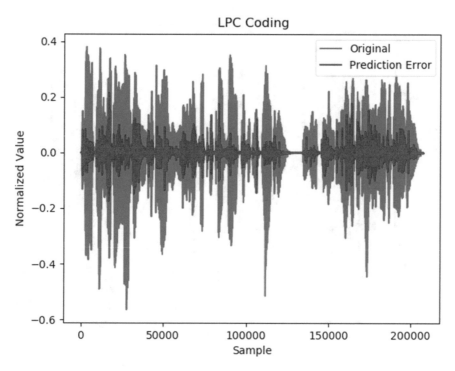

Figure 3.3: Plot of the original speech waveform and the prediction error of the LPC predictor. Observe that the prediction error is indeed much smaller than the original.

As a test signal we use "fspeech.wav", which is a female speech signal at 32 kHz sampling rate. Execute the example with

```
python3 lpcexample.py
```

and listen to the original speech sound, then the prediction error, look at the plots of the waveforms for the original and the prediction error, Fig. 3.3, and after closing that plot listen to the reconstructed speech signal. For this signal the program computes and displays the Signal to Prediction Error Power ratio as 61.54235163995218, or $10 \cdot \log_{10}(61.54) \approx 18$ dB. This is also called the "prediction gain". We cannot use this directly for coding/decoding, because we still have to include quantization (Sect. 3.3.3). Observe that the signal reconstructed from the prediction error is indeed identical to the original (Fig. 3.4).

3.3.2 Least Mean Squares (LMS) Adaptation

The LPC coder has the advantage that it is simple and efficient, and it has a very simple decoder, but the disadvantage that the encoder needs to have a delay of at least 1 signal block to compute the prediction coefficients, and that it needs to transmit the prediction coefficients as side information, which means we should keep its prediction order small in order to limit the number of bits for this side information.

LMS adaptation takes a different approach. Instead of computing a completely new set of prediction coefficients for each block, it updates the prediction coefficients for each new arriving sample. Since this is based on past (reconstructed) samples, the same

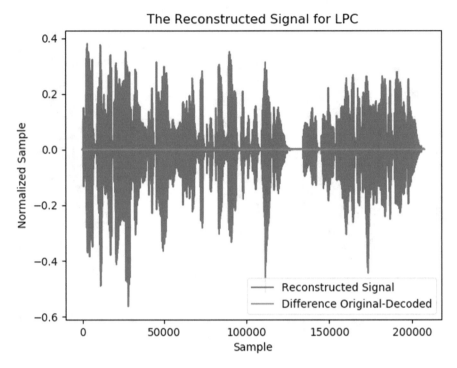

Figure 3.4: Plot of the reconstructed speech waveform after decoding and the reconstruction error of the LPC predictor. Observe that the reconstruction error is practically zero.

computation can also be done in the decoder, which means we do not need to transmit them as side information, and we have no signal delay! The disadvantage is that the decoder becomes more computationally complex than in the LPC case, since it also needs to compute the prediction coefficients.

To obtain the update for the prediction coefficients, an algorithm known from optimization called "Stochastic Gradient Descent" (SGD) [8] is applied. We want to minimize the expectation of the squared prediction error $E(e^2(n))$ as a function of our prediction coefficients $\mathbf{h}$. SGD takes many steps in the direction of the gradient of this function, to approach its minimum. Since many steps are taken anyway, it is assumed that the averaging of the Expectation is done implicitly during these many steps. Hence SGD simplifies the function to minimize by removing the expectation step from it. This is practical for an implementation because we do not need to compute an explicit averaging of $e^2(n)$, which reduces the computational complexity. Now the function to minimize in each step is just the current squared prediction error. To show that it also depends on the prediction coefficients, we write its vector as additional argument,

$$f(n, \mathbf{h}(n)) := e^2(n, \mathbf{h}(n)) := \left(x(n) - \sum_{m=0}^{L-1} x(n-1-m) \cdot h_m(n) \right)^2 \qquad (3.9)$$

To denote the time dependence of our predictor coefficients, we write the vector of its coefficients as $\mathbf{h}(n)$ and its components as $h_m(n)$, where m is the position inside the

116

impulse response $\mathbf{h}(n)$. The gradient of our error function $f(n, \mathbf{h}(n))$ with respect to $\mathbf{h}(n)$ then becomes the vector

$$\nabla f(n, \mathbf{h}) = \left[\frac{\delta f(n, \mathbf{h}(n))}{\delta h_0(n)}, \cdots, \frac{\delta f(n, \mathbf{h}(n))}{\delta h_{L-1}(n)} \right]$$

Looking at Eq. (3.9) and using the chain rule for the derivation, we can see that this simplifies to

$$\nabla f(n, \mathbf{h}(n)) = -2 \cdot e(n) \left[x(n-1), \cdots x(n-L) \right] \tag{3.10}$$

The SGD update step according to [8] is

$$\mathbf{h}(n+1) = \mathbf{h}(n) - \eta \cdot \nabla f(n, \mathbf{h}(n))$$

with some step size (or learning rate) η, a tuning parameter, which means it needs to be "tuned" or tried out for the particular problem. In our case, using Eq. (3.10), it simplifies to

$$\mathbf{h}(n+1) = \mathbf{h}(n) + \mu \cdot e(n) \cdot \left[x(n-1), \cdots x(n-L) \right],$$

where $\mu := 2 \cdot \eta$, or with

$$\mathbf{x}(n) := \left[x(n-1), \cdots x(n-L) \right]$$

this simply becomes

$$\mathbf{h}(n+1) = \mathbf{h}(n) + \mu \cdot e(n) \cdot \mathbf{x}(n)$$

This is now the famous Least Mean Squares (LMS) filter according to Widrow–Hoff [3, 9–12]. This has the slight drawback that the choice of the tuning parameter μ depends on the scaling of the signal $x(n)$, for instance, if it is normalized between -1 and 1, or a 16 bit integer value between $-32{,}768$ and $+32{,}767$, and it also depends on the strength of the signal within this range. To make it less dependent on these factors, the "Normalized LMS" (NLMS) was introduced [13], which basically normalizes the update term to the current signal power $\mathbf{x}(n) \cdot \mathbf{x}^T(n)$,

$$\mathbf{h}(n+1) = \mathbf{h}(n) + \mu \cdot e(n) \cdot \frac{\mathbf{x}(n)}{\mathbf{x}(n) \cdot \mathbf{x}^T(n)}$$

To avoid division by zero or very small numbers, often the following formula is used:

$$\mathbf{h}(n+1) = \mathbf{h}(n) + \mu \cdot e(n) \cdot \frac{\mathbf{x}(n)}{a + \mathbf{x}(n) \cdot \mathbf{x}^T(n)} \tag{3.11}$$

with some small positive constant a, which should be chosen much smaller than the expected energy of $\mathbf{x}(n) \cdot \mathbf{x}^T(n)$. For the NLMS algorithm the step size parameter μ is usually chosen between 0 and 1.0, often, simply $\mu = 1.0$ is a good choice.

3.3.3 Quantization, Decoder in Encoder

In an Encoder/Decoder application, the predictors of the encoder and the decoder need to compute the exact same results. The (N)LMS algorithm updates its prediction coefficients with each sample. Hence even tiny differences between encoder/decoder can lead to diverging or drifting prediction coefficients, and may lead to unstable reconstruction filters in the decoder, because that is an IIR filter. To avoid or mitigate this effect, the so-called decoder in encoder structure is used, where a quantized prediction error is not only transmitted to the decoder, but is also used by an internal decoder to reconstruct the past samples on which the predictor works. This ensures that both predictors work on the same data, from the quantized prediction error.

Assume we use a Mid-Tread quantizer and de-quantizer according to Eqs. (2.1), (2.2) with quantization step size Δ for the prediction error. For the encoder Eq. (3.1) we combine both quantizer and de-quantizer into the following:

$$e(n) = x(n) - \hat{x}(n) \tag{3.12}$$

$$e_q(n) := round(e(n)/\Delta) \cdot \Delta \tag{3.13}$$

For its predictor Eq. (3.2) the encoder now uses the internal decoder and the quantized prediction error,

$$x_{rec}(n) = e_q(n) + \hat{x}(n) \tag{3.14}$$

and uses the reconstructed samples $x_{rec}(n)$ as basis for its prediction, as the decoder will also do.

To see the effect of quantization on the reconstruction error $x(n) - x_{rec}(n)$ we rewrite the quantized prediction error as a sum of the unquantized prediction error plus a quantization error $q(n)$,

$$e_q(n) = e(n) + q(n),$$

where $-0.5\Delta \leq q(n) \leq 0.5\Delta$. Then using Eqs. (3.12, (3.13), and (3.14) we obtain

$$x(n) - x_{rec}(n) = x(n) - (e_q(n) + \hat{x}(n)) = x(n) - (e(n) + q(n) + \hat{x}(n))$$
$$= x(n) - ((x(n) - \hat{x}(n)) + q(n) + \hat{x}(n)) = q(n)$$

This means we expect the reconstructed signal after decoding $x_{rec}(n)$ to have the same quantization error as the quantized prediction error signal $e_q(n)$ [6] (Figs. 3.5 and 3.6).

3.3.4 Python Example LMS

The following shows a Python example for an NLMS encoder with quantization and a decoder in encoder structure. A step size of $\mu = 1.0$ and $a = 0.1$ is chosen.

```
x, fs = wavread('fspeech.wav')
#normalized float, −1<x<1
x = np.array(x,dtype=float)/2**15
```

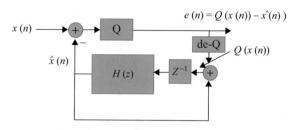

Figure 3.5: A block diagram of a predictive encoder, with quantization and a decoder in encoder structure. Observe the reconstruction with the feedback path of the prediction error signal.

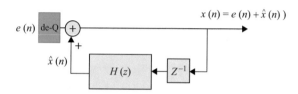

Figure 3.6: A block diagram of a predictive decoder, with de-quantization.

```
sound(2**15*x, fs)
print("np.size(x)=", np.size(x))
e = np.zeros(np.size(x))
xrek=np.zeros(np.size(x));
P=0;
L=10
h = np.zeros(L)
#have same 0 starting values as in decoder:
x[0:L]=0.0
quantstep=0.01;
#Encoder:
for n in range(L, len(x)):
    if n> 4000 and n< 4002:
        print("encoder_h:_", h)
    #prediction error and filter, using the vector of reconstructed samples:
    #predicted value from past reconstructed values:
    xrekvec=xrek[n−L+np.arange(L)]
    P=np.dot(np.flipud(xrekvec), h)
    #quantize and de−quantize e to step−size 0.05 (mid tread):
    e[n]=np.round((x[n]−P)/quantstep)*quantstep;
    #Decoder in encoder, new reconstructed value:
    xrek[n]=e[n]+P;
    #LMS update rule:
    #h = h + 1.0* e[n]*np.flipud(xrekvec)
    #NLMS update rule:
    h = h + 1.0* e[n]*np.flipud(xrekvec)/(0.1+np.dot(xrekvec,xrekvec))
```

At this point we can compute the signal power to prediction error power ratio,

print("The_Signal_to_Error_ratio_is:", np.dot(x.transpose(),x)/np.dot(e.transpose(),e))
#The Signal to Error ratio is: 28.479576824 for LMS.
#The Signal to Error ratio is: 39.35867161114023 for NLMS.

For this speech signal we get a Signal to Prediction Error ratio of about 39, or $10 \cdot \log 10(39) \approx 16\,\text{dB}$. The higher this ratio the better we can compress. As a rough estimate according to Eq. (2.6) (1 bit for 6 dB SNR), with this dB value we could save about 2.5 bits per sample. This is less than for the LPC example in Sect. 3.3.1, but here we now included quantization, such that we can really use it in an encoder/decoder.

The corresponding decoder is

```
h = np.zeros(L);
xrek = np.zeros(np.size(x));
for n in range(L, len(x)):
    if n> 4000 and n< 4002:
        print("decoder_h:_", h)
    P=np.dot(np.flipud(xrek[n−L+np.arange(L)]), h)
    xrek[n] = e[n] + P
    xrekvec=xrek[n−L+np.arange(L)]
    #LMS update:
    #h = h + 1.0 * e[n]*np.flipud(xrekvec);
```

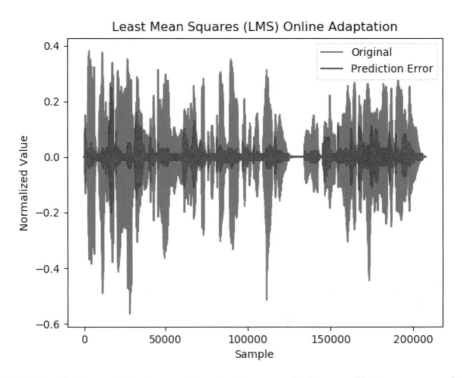

Figure 3.7: Plot of the original speech waveform and the prediction error of the LMS predictor with quantization. Observe that the prediction error, the difference Original-Decoded, is also much smaller than the original.

#NLMS update:

h = h + 1.0* e[n]*np.flipud(xrekvec)/(0.1+np.dot(xrekvec,xrekvec))

We can execute the example from our terminal shell with
`python3 lmsquantexample.py`
In the beginning it loads and plays a speech sound file, then it computes and plots the prediction error together with the original signal, Fig. 3.7, then it computes the decoder, plots the reconstructed speech file and the reconstruction error, Fig. 3.8, and plays the reconstructed speech signal. Observe that the reconstruction error is of the size of the quantization error for the prediction error. The "print" statements of the prediction filter coefficients "h" show that they are indeed identical between encoder and decoder.

3.3.5 Prediction for Lossless Coding

If our audio signal consists of integer valued samples, for instance, the 16 bit integers most often used, we can easily use predictive coding to obtain a lossless coder (despite quantization), by simply rounding the predictor output to the nearest integer as quantization. This is the same as using this quantization with our decoder in encoder structure of Sect. 3.3.3, but since the input signal is now also integer valued, it simplifies the structure. This approach maps the integer valued input signal $x(n)$ to an integer valued prediction error $e_q(n)$, in a reversible way, which makes it lossless.

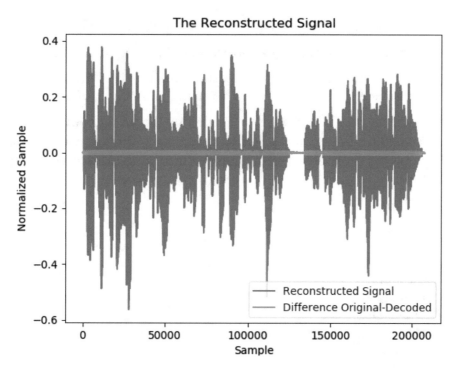

Figure 3.8: Plot of the reconstructed speech waveform after decoding and the reconstruction error of the LMS predictor with quantization. Observe that the reconstruction error is of the size of the quantization step size.

Our encoder equation (3.1) then becomes

$$e_q(n) := x(n) - round(\hat{x}(n)). \qquad (3.15)$$

and the decoder equation (3.2) becomes

$$x_{rec}(n) = e(n) + round(\hat{x}(n)) \qquad (3.16)$$

3.3.6 Python Example for Predictive Lossless Coding

The following shows a Python example for lossless coding using NLMS prediction. The encoder is,

```
for n in range(L, len(x)):
    if n> 4000 and n< 4002:
        print("encoder_h:_", h)
    #prediction error and filter, using the vector of reconstructed samples,
    #predicted value from past reconstructed values, since it is lossless, xrek=x:
    xrekvec=x[n−L+np.arange(L)]
    P=np.dot(np.flipud(xrekvec), h)
    #quantize and de−quantize by rounding to the nearest integer:
    P=round(P)
    #prediction error:
    e[n]=x[n]−P
    #NLMS update:
    h = h + 1.0* e[n]*np.flipud(xrekvec)/(0.1+np.dot(xrekvec,xrekvec))
```

At this point we can compute the signal power to prediction error power ratio,

```
print("The_Signal_to_Error_ratio_is:", np.dot(x.transpose(),x)/np.dot(e.transpose(),e))
#The Signal to Error ratio is: 27.79929537470223, a little less than with quant for NLMS.
```

For this speech signal we get a Signal to Prediction Error ratio of about 28, or $10 \cdot \log 10(28) \approx 14\,\text{dB}$. This is somewhat less than for the LMS case with quantization in Sect. 3.3.4, but now we have obtained lossless coding. It is a prediction gain of about 2.3 bits per sample.

The decoder is,

```
# Decoder
h = np.zeros(L);
xrek = np.zeros(np.size(e));
for n in range(L, len(x)):
    if n> 4000 and n< 4002:
        print("decoder_h:_", h)
    xrekvec=xrek[n−L+np.arange(L)]
    P=np.dot(np.flipud(xrekvec), h)
    P=round(P)
    xrek[n] = e[n] + P
    #NLMS update:
    h = h + 1.0* e[n]*np.flipud(xrekvec)/(0.1+np.dot(xrekvec,xrekvec))
```

We can execute the example from our terminal shell with
`python3 lmslosslessexample.py`
In the beginning it again loads and plays a speech sound file, then it computes and plots and plays the prediction error together with the original signal, Fig. 3.9, then it computes the decoder, plots the reconstructed speech file and the reconstruction error, Fig. 3.10, and plays the reconstructed speech signal. We can see that the prediction error is indeed zero. The "print" statements of the prediction filter coefficients "h" show that they are again indeed identical between encoder and decoder.

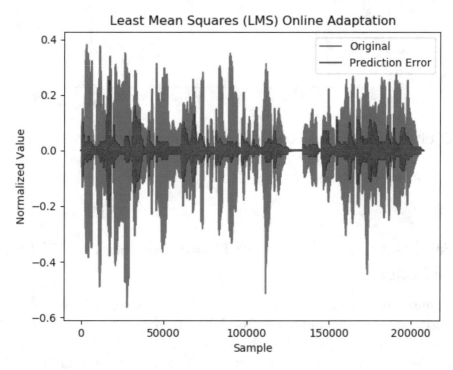

Figure 3.9: Plot of the original speech waveform and the prediction error of the lossless LMS predictor. Observe that the prediction error, the difference Original-Decoded, is also much smaller than the original.

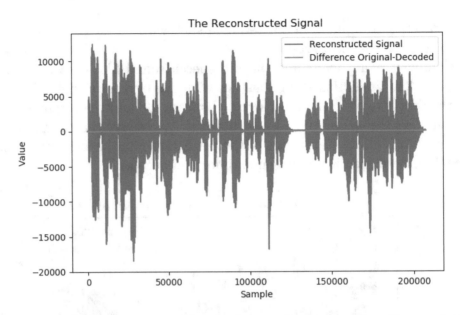

Figure 3.10: Plot of the reconstructed speech waveform after decoding and the reconstruction error of the LMS lossless predictor. Observe that the reconstruction error is indeed zero.

3.3.7 Weighted Cascaded Least Means Squares (WCLMS) Prediction

To obtain improved prediction accuracy, Weighted Cascaded Least Means Squares (WCLMS) uses several LMS predictors to obtain an improved prediction performance [14]. Assume predictors P_1, P_2, P_3 with different orders for different signal statistics. How can they be combined? We use predictive "Minimum Description Length" principle [15] for the "optimal" combination of predictors. We combine the predictors as a weighted average, where the weights w_i are determined by the probability of the corresponding predictor of being "correct" on the past signal,

$$P(n) = \sum_{i=1}^{3} w_i(n) \cdot P_i(n) \tag{3.17}$$

Assume that the prediction error has a Laplacian distribution [16],

$$p(x) = \frac{c}{2} \cdot e^{-c \cdot |x|}.$$

We can compute each predictors past prediction error over some window $h_W(n)$ with length N as

$$e_i(n) = \sum_{m=1}^{N} h_W(n) \cdot |x(m-n) - P_i(n)|.$$

The weights $w_i(n)$ are then

$$w_i(n) = k \cdot e^{-c \cdot e_i(n)},$$

where c depends on the error distribution and constant k is chosen such that the weights sum up to one, $\sum_{i=1}^{3} w_i(n) = 1$.

We choose these three predictors to result from a cascade of predictors. The first in the cascade predicts the audio signal, and the next two predict their previous predictor's prediction error,

$$\hat{x}(n) = \sum_{m=0}^{L_1-1} x(n-1-m) \cdot h_1(m)$$
$$\times e_1(n) = x(n) - \hat{x}(n)$$
$$\hat{e_1}(n) = \sum_{m=0}^{L_2-1} e_1(n-1-m) \cdot h_2(m)$$
$$\times e_2(n) = e_1(n) - \hat{e_1}(n)$$
$$\hat{e_2}(n) = \sum_{m=0}^{L_3-1} e_2(n-1-m) \cdot h_3(m)$$

Each predictor is updated using the NLMS method of Eq. (3.11). We then obtain our three predictors by adding the predicted prediction error in each consecutive stage as

$$P_1(n) = \hat{x}(n)$$
$$P_2(n) = \hat{x}(n) + \hat{e_1}(n)$$
$$P_3(n) = \hat{x}(n) + \hat{e_1}(n) + \hat{e_2}(n)$$

These predictions are then used in the WCLMS prediction equation (3.17). An example for the WCLMS predictor is the file `wclms_losslessexample.py`. There, WCLMS is implemented with $L_1 = 30$, $L_2 = 20$, $L_3 = 10$, and $c = 0.01$. For this and again our "fspeech.wav" example we obtain a Signal to Error Power ratio of 37.3, or 15.7 dB, which is almost 2 dB better than our NLMS lossless example of Sect. 3.3.6 with predictor length $L = 10$.

References

1. *Moore–Penrose inverse*, https://en.wikipedia.org/wiki/Moore. Accessed Jan 2019

2. *Toeplitz matrix*, https://en.wikipedia.org/wiki/Toeplitz_matrix. Accessed Dec 2018

3. M. Hayes, *Statistical Signal Processing and Modelling* (Wiley, London, 1996)

4. B. Gold, N. Morgan, *Speech and Audio Signal Processing* (Wiley, London, 2000)

5. J.R. Deller, Jr., J.G. Proakis, J.H. Hansen, *Discrete-Time Processing of Speech Signals* (Macmillan, New York, 1993)

6. N.S. Jayant, P. Noll, *Digital Coding of Waveforms: Principles and Applications to Speech and Video* (Prentice Hall, Englewood, 1990)

7. A. Spanias, T. Painter, V. Atti, *Audio Signal Processing and Coding* (Wiley, London, 2005)

8. *Stochastic gradient descent*, https://en.wikipedia.org/wiki/Stochastic_gradient_descent. Accessed Jan 2019

9. *Least mean squares filter*, https://en.wikipedia.org/wiki/Least_mean_squares_filter. Accessed Jan 2019

10. B. Widrow, M.E. Hoff, Adaptive switching circuits, in *WESCON Conv. Rec* (1960)

11. B. Widrow, S.D. Stearns, *Adaptive Signal Processing* (Prentice Hall, 1985)

12. J.G. Proakis, C.M. Rader, F. Ling, C.L. Nikias, *Advanced Digital Signal Processing* (Macmillan, New York, 1992)

13. S. Haykin, *Adaptive Filter Theory* (Prentice Hall, Englewood, 2002)

14. G.D.T. Schuller, B. Yu, D. Huang, B. Edler, Perceptual audio coding using adaptive pre- and post-filters and lossless compression. IEEE Trans. Speech Audio Process. **10**, 379–390 (2002)

15. *Minimum description length*, https://en.wikipedia.org/wiki/Minimum_description_length. Accessed Jan 2019

16. *Laplace distribution*, https://en.wikipedia.org/wiki/Laplace_distribution. Accessed Jan 2019

4 Psycho-Acoustic Models

0	1	2	3	4	5	6	7	8	9	10	11	12
0	100	200	300	400	510	630	770	920	1080	1270	1480	1720
13	14	15	16	17	18	19	20	21	22	23	24	
2000	2320	2700	3150	3700	4400	5300	6400	7700	9500	12,000	15,500	

Table 4.1: A table containing Bark numbers (above) and corresponding frequencies in Hertz (below).

4.1 Introduction

In audio coding we need a psycho-acoustic model for the irrelevance reduction, to avoid encoding information which is not perceived by the receiver, the human ear. Specifically, the psycho-acoustic model is used to control the step size of the quantizers of the audio signal such that the quantization error is (at least ideally) not noticeable. Psycho-acoustic models are not standardized for MPEG audio coders, they are left as proprietary technology of the encoder developers. So the following chapter shows the building blocks of a typical psycho-acoustic model.

4.2 The Bark Scale

The inner ear can be modeled as a filter bank, e.g. a Gammatone filter bank [1, 2], consisting of bandpass filters of corresponding width (but much higher number of bands) of the so-called Bark frequency scale [3–6] s

These frequencies form a function in the form of a table, where these frequencies are numbered, starting with 0. These numbers are then the Bark numbers, as can be seen in Table 4.1.

For audio coding only the basic mapping function of the frequency scale matters, hence for our example implementation we use the simplest approximation for it the Schroeder approximation.

© Springer Nature Switzerland AG 2020
G. Schuller, *Filter Banks and Audio Coding*,
https://doi.org/10.1007/978-3-030-51249-1_4

4.2.1 The Schroeder Approximation

This approximation was first described by M. R. Schroeder in [7]. Later it was used in [8], and it is also used in the PEAQ standard for objective quality estimation [9]

It is the simplest approximation, again with f in Hertz to Bark [7]:

$$z = 6 \cdot \operatorname{arcsinh}(f/600)$$

It has an exact inverse, Bark to Hertz:

$$f = 600 \cdot \sinh(z/6)$$

4.2.2 Bark Scale Mapping

The following shows an example of an implementation of a Bark scale mapping. We choose 64 subbands in the Bark scale, if our audio band covers the 24 Bark bands, this mean each Bark subband is about 1/3 Bark wide.

In Python we construct a matrix W for this mapping, because a matrix multiplication is much faster in Python than a "for" loop. Horizontally it contains our subbands on the linear frequency scale, and vertically the Bark subbands. For each linear subband it contains 1's at the position of the corresponding 1/3 Bark subband:

```
def bark2hz(Brk):
    """ Usage:
    Hz=bark2hs(Brk)
    Args :
        Brk : (ndarray) Array containing Bark scaled values.
    Returns :
        Fhz : (ndarray) Array containing frequencies in Hz.
    """
    Fhz = 600. * np.sinh(Brk/6.)
    return Fhz

def mapping2barkmat(fs, nfilts,nfft):
```

We can visualize this matrix by displaying the first 256 columns as an image, with the following code:

```
ipython --pylab
from psyacmodel import *
W=mapping2barkmat(fs=32000,nfilts=64,nfft=2048)
```

```
imshow(W[:,:256],cmap='Blues')
title('Matrix W as Image')
xlabel('Uniform Subbands')
ylabel('Bark Subbands')
```

For the actual mapping from the magnitude of complex uniformly spaced DFT sub-bands for the psycho-acoustic model to the Bark subbands, we add the signal **powers** from the corresponding DFT bands. Then we take the square root to obtain a "voltage" again. As Python function:

def mapping2bark(mX,W,nfft):
    #Maps (warps) magnitude spectrum vector mX from DFT to the Bark scale
    #arguments: mX: magnitude spectrum from fft
    #W: mapping matrix from function mapping2barkmat
    #nfft: : number of subbands in fft
    #returns: mXbark, magnitude mapped to the Bark scale
    nfreqs=**int**(nfft/2)
    #Here is the actual mapping, suming up powers and conv. back to Voltages:
    mXbark = (np.dot(np.**abs**(mX[:nfreqs])**2.0, W[:, :nfreqs].T))**(0.5)
    return mXbark

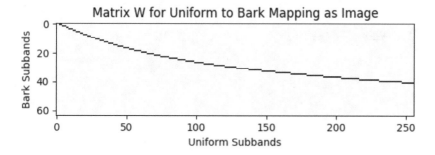

Figure 4.1: Our mapping matrix from Hertz to Bark as image, for the first 256 subbands (blue means 1).

4.2.3 Mapping from Bark Scale Back to Uniform

After having computed the masking threshold in the Bark scale, we need to map it back to the uniform scale of our filter bank. For that we need to "distribute" the corresponding power of each of our Bark subbands into the corresponding filter bank bands on the linear frequency scale. Then we take the square root to obtain a "voltage" again.

We again construct a matrix to do that in Python. When there is one uniform subband in the Bark subband, it gets a factor 1, if there are 2 subbands, they get a factor of sqrt(2), and so on, using a diagonal matrix multiplication for those factors. It is an 64×1024 matrix.

def mappingfrombarkmat(W,nfft):

  #Constructing inverse mapping matrix W_ inv from matrix W for mapping back from bark scale

  #usuage: W_ inv=mappingfrombarkmat(Wnfft)

  #argument: W: mapping matrix from function mapping2barkmat

  #nfft: : number of subbands in fft

  nfreqs=**int**(nfft/2)

  W_inv= np.dot(np.diag((1.0/(np.**sum**(W,1)+1e−6))**0.5), W[:,0:nfreqs + 1]).T

We can plot the first 256 rows of this matrix `W_inv` as an image,

```
ipython --pylab
from psyacmodel import *
W_inv=mappingfrombarkmat(W,nfft=2048)
imshow(W_inv[:256,:],cmap='Blues')
title('Matrix W_inv as Image')
xlabel('Bark Subbands')
ylabel('Uniform Subbands')
```

Matrix W_inv for Bark to Uniform Mapping as Image

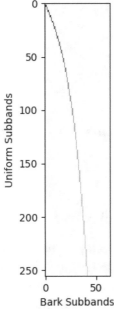

Figure 4.2: Our mapping matrix from Bark to Hertz as image, for the first 256 uniform subbands (blue means 1).

4.3 Hearing Threshold in Quiet

The hearing threshold in quiet is a curve which denotes the boundary of audibility of sound for the ear in quiet, at different frequencies. A commonly used approximation formula can be found in [10, 11]. For f: frequency in Hz and LTQ: the masking threshold in quiet on a dB sound pressure scale, it is,

$$LTQ = 3.64 \cdot (f/1000)^{-0.8} - 6.5 \cdot e^{-0.6 \cdot (f/1000 - 3.3)^2} + 10^{-3} \cdot (f/1000)^4$$

The resulting function is plotted in Fig. 4.3. It can be seen that the maximum sensitivity of the ear in quiet is in the range of 3–4 kHz. But keep in mind that this curve as an average, there is a considerable deviation from this curve among individuals, and with age this threshold increases for higher frequencies.

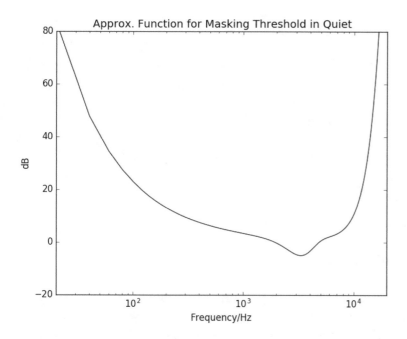

Figure 4.3: The approximation function for the hearing threshold in quiet.

On the Bark scale we can compute the threshold in quiet with the following code:

```
#Threshold in quiet:
maxfreq=fs/2.0
maxbark=hz2bark(maxfreq)
step_bark = maxbark/(nfilts-1)
barks=np.arange(0,nfilts)*step_bark
#convert the bark subband frequencies to Hz:
f=bark2hz(barks)+1e-6
#Threshold of quiet in the Bark subbands in dB:
LTQ=np.clip((3.64*(f/1000.)**-0.8 -6.5*np.exp(-0.6*(f/1000.-3.3)**2.)
    +1e-3*((f/1000.)**4.)),-20,120)
```

4.4 The Spreading Function

Now that we have the mappings to and from the Bark scale, we can apply the spreading function of psycho-acoustics, which is defined in the Bark scale. The spreading function describes the elevation of the masking threshold at the presence of a sinusoidal signal, a pure tone, or (narrowband) noise (see also [10, 12]). It depends on its sound pressure level L_s and its frequency f. The sound pressure level is the sound pressure relative to (in this case) the sound pressure at the masking threshold at the highest sensitivity of the ear, in dB. If we decompose an audio signal into its sinusoidal components, then each of those has its associated spreading function. On the Bark scale, the spreading function is a triangular curve which is centred at its peak at some level below the sound pressure level of the tone. This level below the tone is called "simultaneous masking", $O_f(z)$, with z being the frequency of the tone f in Bark. It is (according to [11])

$$O_f(z)/\text{dB} = \alpha \cdot (14.5 + z) + (1 - \alpha) \cdot \alpha_\nu$$

, where α is a tonality coefficient ($\alpha = 1$ for tonal signals, $\alpha = 0$ for noise-like signals). α_ν is the "noise coefficient", which [13] approximates to $\alpha_\nu = 5.5\,\text{dB}$. This results in

$$O_f(i)/\text{dB} = \alpha \cdot (14.5 + z) + (1 - \alpha) \cdot 5.5$$

The lower slope of the triangular shaped spreading function has an inclination of

$$S_1 = 27\,\text{dB/Bark}.$$

Its upper slope has an inclination of

$$S_2 = (24 + 0.23/(\text{f/kHz}) - 0.2\,L_{\text{s/dB}})\text{dB/Bark}$$

Terhardt [10] and Zoelzer [11]. In practice, for our model implementation, we see that we can make further simplifying assumptions. We assume a tonal signal ($\alpha = 1$), a sound pressure level of $L_s = 60\,\text{dB}$ (approximately the sound pressure level of speech in a conversation), a tone frequency of $f = 1\,\text{kHz}$ or approximately $z = 9$ Bark. This leads to

$$O_f(z) \approx 23.5\,\text{dB}$$
$$S_1 = 27\,\text{dB/Bark}$$
$$S_2 = (24 + 0.23/(\text{f/kHz}) - 0.2L_{\text{s/dB}})\text{dB/Bark} \approx 12\,\text{dB/bark}$$

For our simplification we define a spreading function prototype, centred at frequency zero and with magnitude 1 as $f_{SPdB}(z)$, which we apply at every frequency and tone magnitude, by shifting it at the frequency z_t of each tone of our mixture, and multiply it with the tones magnitude $x_M(z_t)$, to obtain each tones spreading function approximation.

We can construct this prototype spreading function as an array in Python. We like to avoid negative indices, hence we place its centre in the middle of the array (instead of zero), at 22 Bark, and shift it from there as needed. We use the following function f_SP_dB, which we store in "psyacmodel.py". The argument `nfilts` is the number of subbands in the Bark domain, in our examples 64, which we use for the resolution of our psycho-acoustic model,

def f_SP_dB(maxfreq,nfilts):
    #usage: spreadingfunctionmatdB=f_SP_dB(maxfreq,nfilts)
    #computes the spreading function protoype, in the Bark scale.
    #Arguments: maxfreq: half the sampling freqency
    #nfilts: Number of subbands in the Bark domain, for instance 64
    maxbark=hz2bark(maxfreq) *#upper end of our Bark scale:22 Bark at 16 kHz*
    #Number of our Bark scale bands over this range: nfilts=64
    spreadingfunctionBarkdB=np.zeros(2*nfilts)
    #Spreading function prototype, "nfilts" bands for lower slope
    spreadingfunctionBarkdB[0:nfilts]=np.linspace(−maxbark*27,−8,nfilts)−23.5
    # "nfilts" bands for upper slope:
    spreadingfunctionBarkdB[nfilts:2*nfilts]=np.linspace(0,−maxbark*12.0,nfilts)−23.5
    return spreadingfunctionBarkdB

If we convert this function in dB back to a power ratio, we call it the intensity of the spreading function $f_{SP}(z)$. To obtain the spreading function for the k'th sinusoid or tone, we multiply its magnitude with the spreading function prototype function, with its peak placed on the frequency of the tone,

$$I_{T,k}(z) = x_M(z_t)f_{SP}(z - z_t).$$

We can plot the spreading function for $z_t = 13$ Bark (for a tone with magnitude 0 dB) on the dB scale, with $maxfreq = 16,000$ or 22 Bark, with

```
ipython --pylab
from psyacmodel import f_SP_dB
spreadingfunctionBarkdB=f_SP_dB(maxfreq=16000,nfilts=64)
maxbark=hz2bark(16000)
#x-axis: 24 Bark in 64 steps:
bark=linspace(0,maxbark,64)
#The prototype over "nfilt" bands or 24 Bark, its center
#shifted down to 24-26/nfilts*24=14.25 Bark:
plot(bark,spreadingfunctionBarkdB[26:90])
axis([5,22,-100,0])
xlabel('Bark')
ylabel('dB')
title('Spreading Function')
```

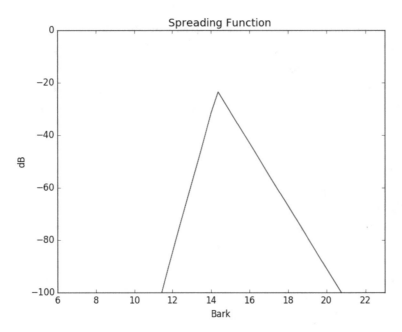

Figure 4.4: Part of a spreading function $I_{T,k}(z)$ for a sinusoid or tone near 14 Bark on the Bark scale, in dB.

In Fig. 4.4 we can see part of this prototype spreading function. At its tip we see the simultaneous masking at -23.5 dB, we see the lower slope towards the lower frequencies at 27 dB/Bark, and the upper slope towards the higher frequencies with 12 dB/Bark.

4.5 Non-linear Superposition

Since our audio signal usually consists of many sinusoidal components, each with its associated spreading function, the question is, how do they combine into an overall masking function? Many psycho-acoustic models use the addition of equivalent powers of the spreading function (an exponent of 2) for the superposition of the spreading functions of many components of an audio signal. For this type of superposition, the distinction of tonal and non-tonal parts becomes important. Non-linear superposition uses exponents different from 2 to obtain a model which does not need a tonality estimation, according to [14] and [15]. We obtain the overall masking threshold $I_T(z)$ from the superposition of the spreading functions $I_{T,k}(z)$ as

$$I_T(z) = \left(\sum_k I_{T,k}(z)^a \right)^{1/a}$$

with k the index for the sinusoids or tones in our audio signal, and z the Bark scale variable. According to the references, $a = 0.3$ is in good agreement with psycho-acoustics. This exponent a is for powers, but in our implementations we use equivalents of voltages instead, hence we have to multiply this exponent by 2, such that we obtain $a = 0.6$.

Experimentally we found that $a = 0.8$ works better, hence this is what we are using in the following examples.

For simplicity of the implementation we assume one sinusoidal component per Bark subband, and that the values in our array mXbark from Sect. 4.2.2 represent the amplitudes of those sinusoids. For the superposition we convert our spreading function from dB to "voltage",

#nfilts: Number of subbands in the Bark domain, for instance 64

For the superposition we construct a matrix, where each Bark subband has a correspondingly shifted spreading function, with the function `spreadingfunctionmat`,

```
def spreadingfunctionmat(spreadingfunctionBarkdB,alpha,nfilts):
    #Turns the spreading prototype function into a matrix of shifted versions.
    #Convert from dB to "voltage" and include alpha exponent
    #nfilts: Number of subbands in the Bark domain, for instance 64
    spreadingfunctionBarkVoltage=10.0**(spreadingfunctionBarkdB/20.0*alpha)
    #Spreading functions for all bark scale bands in a matrix:
    spreadingfuncmatrix=np.zeros((nfilts,nfilts))
    for k in range(nfilts):
        spreadingfuncmatrix[k,:]=spreadingfunctionBarkVoltage[(nfilts-k):(2*nfilts-k)]
    return spreadingfuncmatrix
```

The non-linear superposition is then implemented using matrix multiplication (again to avoid slow "for" loops) in the function `maskingThresholdBark`. It also includes the hearing threshold in quiet in the Bark domain, such that we can obtain the maximum between it and the spreading functions. The functions output is the masking threshold for the audio signal in the Bark domain.

```
def maskingThresholdBark(mXbark,spreadingfuncmatrix,alpha,fs,nfilts):
    #Computes the masking threshold on the Bark scale with non-linear superposition
    #usage: mTbark=maskingThresholdBark(mXbark,spreadingfuncmatrix,alpha)
    #Arg: mXbark: magnitude of FFT spectrum, on the Bark scale
    #spreadingfuncmatrix: spreading function matrix from function spreadingfunctionmat
    #alpha: exponent for non-linear superposition (eg. 0.6),
    #fs: sampling freq., nfilts: number of Bark subbands
    #nfilts: Number of subbands in the Bark domain, for instance 64
    #Returns: mTbark: the resulting Masking Threshold on the Bark scale

    #Compute the non-linear superposition:
    mTbark=np.dot(mXbark**alpha, spreadingfuncmatrix**alpha)
    #apply the inverse exponent to the result:
    mTbark=mTbark**(1.0/alpha)
    #Threshold in quiet:
    maxfreq=fs/2.0
    maxbark=hz2bark(maxfreq)
    step_bark = maxbark/(nfilts-1)
    barks=np.arange(0,nfilts)*step_bark
    #convert the bark subband frequencies to Hz:
```

```
f=bark2hz(barks)+1e−6
#Threshold of quiet in the Bark subbands in dB:
LTQ=np.clip((3.64*(f/1000.)**−0.8 −6.5*np.exp(−0.6*(f/1000.−3.3)**2.)
    +1e−3*((f/1000.)**4.)),−20,120)
#Maximum of spreading functions and hearing threshold in quiet:
mTbark=np.max((mTbark, 10.0**((LTQ−60)/20)),0)
return mTbark
```

4.6 The Complete Psycho-Acoustic Model

Now we can put our functions together to a complete model. It starts with taking the magnitude of the DFT of our input **x**, because a complex spectrum gives us a more precise magnitude estimate for tones. Then it applies the mapping to the Bark scale, followed by applying the spreading functions, their non-linear superposition, and the hearing threshold in quiet in the function `maskingThresholdBark`. Then it maps the resulting masking threshold back to the linear frequency domain,

```
mX=np.abs(np.fft.fft(x[0:2048],norm='ortho'))[0:1025]
mXbark=mapping2bark(mX,W,nfft)
#Compute the masking threshold in the Bark domain:
mTbark=maskingThresholdBark(mXbark,spreadingfuncmatrix,alpha,fs,nfilts)
#Massking threshold in the original frequency domain
```

To listen to test sounds, we use the function "sound" in a file "sound.py", to play back sound to our sound device, using the "pyaudio" library,

```
def sound(audio, samplingRate):
  #funtion to play back an audio signal, in array "audio"
    import pyaudio
    if len(audio.shape)==2:
        channels=audio.shape[1]
        print("Stereo")
    else:
        channels=1
        print("Mono")
    p = pyaudio.PyAudio()
    # open audio stream

    stream = p.open(format=pyaudio.paInt16,
                channels=channels,
                rate=samplingRate,
                output=True)

    #Clipping to avoid overloading the sound device:
    audio=np.clip(audio,−2**15,2**15−1)
    sound = (audio.astype(np.int16).tostring())
    stream.write(sound)
```

```
    # close stream and terminate audio object
    stream.stop_stream()
    stream.close()
    p.terminate()
    return
```

We can now illustrate our psycho-acoustic model with the main routine in our file "psy-acmodel.py",

```
if __name__ == '__main__':
    #testing:
    import matplotlib.pyplot as plt
    import sound

    fs=32000 # sampling frequency of audio signal
    maxfreq=fs/2
    alpha=0.8 #Exponent for non−linear superposition of spreading functions
    nfilts=64 #number of subbands in the bark domain
    nfft=2048 #number of fft subbands

    W=mapping2barkmat(fs,nfilts,nfft)
    plt.imshow(W[:,:256],cmap='Blues')
    plt.title('Matrix_W_for_Uniform_to_Bark_Mapping_as_Image')
    plt.xlabel('Uniform_Subbands')
    plt.ylabel('Bark_Subbands')
    plt.show()

    W_inv=mappingfrombarkmat(W,nfft)
    plt.imshow(W_inv[:256,:],cmap='Blues')
    plt.title('Matrix_W_inv_for_Bark_to_Uniform_Mapping_as_Image')
    plt.xlabel('Bark_Subbands')
    plt.ylabel('Uniform_Subbands')
    plt.show()

    spreadingfunctionBarkdB=f_SP_dB(maxfreq,nfilts)
    #x−axis: maxbark Bark in nfilts steps:
    maxbark=hz2bark(maxfreq)
    print("maxfreq=", maxfreq, "maxbark=", maxbark)
    bark=np.linspace(0,maxbark,nfilts)
    #The prototype over "nfilt" bands or 22 Bark, its center
    #shifted down to 22−26/nfilts*22=13 Bark:
    plt.plot(bark,spreadingfunctionBarkdB[26:(26+nfilts)])
    plt.axis([6,23,−100,0])
    plt.xlabel('Bark')
    plt.ylabel('dB')
    plt.title('Spreading_Function')
    plt.show()
```

```
spreadingfuncmatrix=spreadingfunctionmat(spreadingfunctionBarkdB,alpha, nfilts)
plt.imshow(spreadingfuncmatrix)
plt.title('Matrix_spreadingfuncmatrix_as_Image')
plt.xlabel('Bark_Domain_Subbands')
plt.ylabel('Bark_Domain_Subbands')
```

And we can test it with a white noise and a sinusoid audio signal,

```
#- Testing------------------------------------
#A test magnitude spectrum:
# White noise:
x=np.random.randn(32000)*1000
sound.sound(x,fs)

mX=np.abs(np.fft.fft(x[0:2048],norm='ortho'))[0:1025]
mXbark=mapping2bark(mX,W,nfft)
#Compute the masking threshold in the Bark domain:
mTbark=maskingThresholdBark(mXbark,spreadingfuncmatrix,alpha,fs,nfilts)
#Massking threshold in the original frequency domain
mT=mappingfrombark(mTbark,W_inv,nfft)
plt.plot(20*np.log10(mX+1e-3))
plt.plot(20*np.log10(mT+1e-3))
plt.title('Masking_Theshold_for_White_Noise')
plt.legend(('Magnitude_Spectrum_White_Noise','Masking_Threshold'))
plt.xlabel('FFT_subband')
plt.ylabel("Magnitude_('dB')")
plt.show()
#--------------------------------------------
#A test magnitude spectrum, an idealized tone in one subband:
#tone at FFT band 200:
x=np.sin(2*np.pi/nfft*200*np.arange(32000))*1000
sound.sound(x,fs)

mX=np.abs(np.fft.fft(x[0:2048],norm='ortho'))[0:1025]
#Compute the masking threshold in the Bark domain:
mXbark=mapping2bark(mX,W,nfft)
mTbark=maskingThresholdBark(mXbark,spreadingfuncmatrix,alpha,fs,nfilts)
mT=mappingfrombark(mTbark,W_inv,nfft)
plt.plot(20*np.log10(mT+1e-3))
plt.title('Masking_Theshold_for_a_Tone')
plt.plot(20*np.log10(mX+1e-3))
plt.legend(('Masking_Trheshold', 'Magnitude_Spectrum_Tone'))
plt.xlabel('FFT_subband')
plt.ylabel("dB")
```

We execute the main function with
`python psyacmodel.py`.

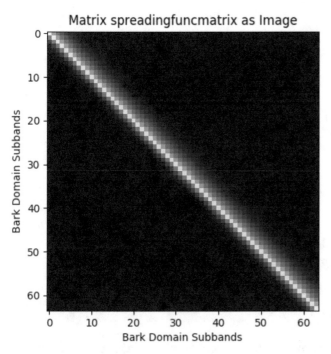

Figure 4.5: The spreading function matrix as image, for its application to the Bark scale subbands.

This program generates Figs. 4.1, 4.2, 4.3, and 4.4, the spreading function matrix as Fig. 4.5, and two audio test signals, a white noise, and a tone, which can be heard while executing the program, and whose masking thresholds are plotted in Figs. 4.6 and 4.7.

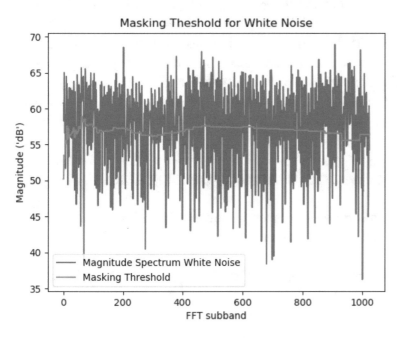

Figure 4.6: The masking threshold for white, spectrally flat noise.

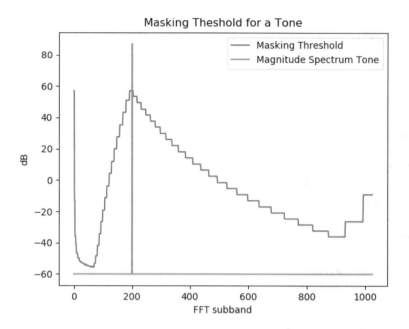

Figure 4.7: The masking threshold for a tone.

4.7 Perceptual Evaluation

For an evaluation or comparison of perceptual coders we need to take the properties of hearing into account. For that reason we obtain the most precise results if we use human test listeners in a listening test. To obtain comparable results, several listening test methods were standardized. An often used standard for "intermediate audio quality" is "MUSHRA", for "Multiple Stimuli with Hidden Reference and Anchor" [16]. A method for the subjective assessment of "small impairments" in audio systems is [17].

In the development process it is desirable to have a (software) tool to quickly measure or estimate the perceptual quality of a coder, or compare different coders or versions of it. A widespread tool for this is "PEAQ", for "Perceptual Evaluation of Audio Quality", which uses psycho-acoustic models to model the process of hearing and in this way aims to predict results of a BS.1116 listening test. It is also standardized as ITU-R Recommendation BS.1387 [18]. A tool which more closely models the physical processes of hearing for the evaluation of coders, and also for evaluating audio separation systems, is "PEMO-Q" [19]. For the assessment of audio source separation it is part of the "PEASS" (Perceptual Evaluation methods for Audio Source Separation) toolbox [20].

4.8 Psycho-Acoustic Models and Quantization

Now that we have a psycho-acoustic model, we need to connect it with the quantization of our subbands, such that the quantization error or "noise" stays inaudible below the masking thresholds.

Our psycho-acoustic model returns the masking thresholds as "voltage" (the square root of the power), and the magnitude of the quantization error needs to stay below our masking threshold "mT". Since the maximum of the magnitude of the quantization error is $delta/2$, we can set

$$delta = 2 * mT.$$

To obtain a precise estimate of the amplitudes of the sinusoids in the signal, the psycho-acoustic model uses the magnitudes of a Short-Time Fourier Transform (see Sect. 1.10.4) instead of the MDCT values, because the MDCT values are real valued, and oscillate between the negative and positive amplitude of the sinusoids in the subbands.

The STFT produces complex values subbands, spread evenly across positive and negative frequencies (the normalized frequency range from $\pi, \ldots 2\pi$ is considered the negative frequency range), where the subband samples of the positive frequencies are conjugate complex to those at the negative frequencies. This means that the magnitudes are identical and we only need the subbands at the positive frequency range of $0, \ldots, \pi$. Our psycho-acoustic model assumes an orthonormal filter bank, which means it fulfils Parseval's theorem for the preservation of energy across the decomposition. It turns out that the Python function `scipy.signal.stft` with its default Hann window and for a block length of 2048 needs a factor of $\sqrt{2048/2}/(2 \cdot 0.375)$ for the analysis to fulfil Parseval's theorem. In order to obtain the same number of STFT subbands in the positive frequency range as our MDCT with N subbands, we choose the STFT block length as $2N$. In our example we choose $N = 1024$.

We define a function `MDCT_psayac_quant_enc`, which contains the MDCT analysis filter bank, the STFT analysis, the psycho-acoustic model and the corresponding quantization, as in an audio encoder. It computes the magnitude of the STFT subbands with

f,t,ys=scipy.signal.stft(x,fs=2*np.pi,nperseg=2*N)
ys *= np.sqrt(2*N/2)/2/0.375

Then it computes the masking thresholds for each block in a "`for`" loop, where we first map the subband magnitudes to the Bark scale with function `mapping2bark`, then compute the masking threshold in the Bark domain with `maskingThresholdBark`, apply logarithmic quantization to the resulting "scalefactors", do their de-quantization, and the mapping back to the linear frequency domain. Here we also introduce a "quality" factor, which allows us to raise or lower the masking threshold. The unchanged masking threshold is obtained with quality = 100, higher values increase the audio quality, lower values decrease the bit rate. Since each Bark subband has its own scalefactor, they are also called "scalefactor bands" [21].

for m **in range**(M): *#M: number of blocks*
  #Observe: MDCT instead of STFT as input can be used by replacing ys by y:
  mXbark=mapping2bark(np.**abs**(ys[:,m]),W,nfft)
  mTbark=maskingThresholdBark(mXbark,spreadingfuncmatrix,alpha,fs,nfilts)/
  (quality/100)
  #Logarithmic quantization of the "scalefactors":
  mTbarkquant[:,m]=np.**round**(np.log2(mTbark)*4) *#quantized scalefactors*
  mTbarkquant[:,m]=np.clip(mTbarkquant[:,m],0,None) *#dequantized is at least 1*
  #Logarithmic de−quantization of the "scalefactors" (decoder in encoder):

```
mTbarkdequant=np.power(2,mTbarkquant[:,m]/4)
#Masking threshold in the original frequency domain
```

Here we can then also try out the difference to simply taking the magnitude of the MDCT subbands instead of the STFT magnitude subbands, by replacing `ys` by `y`.

Next the function quantizes all MDCT subbands using the *delta* from the masking thresholds from our psycho-acoustic model, with uniform mid-tread quantization, and then returns it together with the quantized scalefactors in `mTbarkquant`,

```
#quantization with step size delta,
yq=np.round(y/delta) #uniform mid−tread quantization
```

For the corresponding functions in an audio decoder we define the function `MDCTsyn_dequant_dec`. It computes the de-quantization of the masking threshold in the Bark subbands, the "scalefactors", the quantization step sizes, the de-quantization of the subband values, and MDCT synthesis filter bank,

```
#Logarithmic de−quantization of the masking threshold
#in the Bark subbands, the "scalefactors":
mTbarkdequant=np.power(2,mTbarkquant/4)
#Massking threshold in the original frequency domain
nfft=2*N #number of fft subbands
W=mapping2barkmat(fs,nfilts,nfft)
W_inv=mappingfrombarkmat(W,nfft)
mT=np.transpose(mappingfrombark(np.transpose(mTbarkdequant),W_inv,nfft))
#The quantization step−sizes
delta=mT*2
#delta.shape: 1025,32
delta=delta[:−1,:] #drop the last stft band to obtain an even number of bands
#De−quantization of the subband values:
ydeq=yq*delta
#print("ydeq.shape=", ydeq.shape)
print("Inverse_MDCT")
#MDCT synthesis filter bank in a decoder:
xrek=MDCTsynfb(ydeq,fb)
```

We can then use the function on an audio example. We start with importing our "sound" library, the "scipy.io.wavfile" and "os" libraries, and set the desired number of subbands for the MDCT,

```
#Example, Demo:
import sound
import scipy.io.wavfile as wav
import os
N=1024 #number of MDCT subbands
```

For experimenting, we could also change the number of Bark subbands `nfilts` here (for instance, to 48), to see how it affects the resulting masking threshold and the resulting quantization noise. Then we read a sound ".wav" file (for experimenting we could use any mono sound file here),

#x=x[:,0]

and compute the sine window for the MDCT and call our functions, here with "quality=60" to make distortions more easily visible in the plots and more easily audible in the reconstructed audio,

```
print("Sampling_Frequency=", fs, "Hz")
#Sine window:
fb=np.sin(np.pi/(2*N)*(np.arange(int(1.5*N))+0.5))
print("Encoder_part:")
#MDCT and quantization:
yq, y, mTbarkquant = MDCT_psayac_quant_enc(x,fs,fb,N, nfilts,quality=60)

print("Decoder_part:")
```

Then we can play out the original and reconstructed sounds for comparison,

```
print("Original_Signal")
os.system('espeak_-s_120_"Original_Signal"')
sound.sound(x,fs)
print("Reconstructed_Signal_after_Quantization_according_to_the_Masking_threshold")
os.system('espeak_-s_120_"Reconstructed_Signal_after_Quantization_according_to_the_
    Masking_threshold"')
```

Finally we can plot the quantized scalefactors over the Bark subbands and the quantized subbands over the MDCT subbands, the spectra for one block, and the spectrograms of the original and reconstructed signal,

```
#print("ydeq[3:6,10]=", ydeq[3:6,10])
import matplotlib.pyplot as plt
plt.plot(mTbarkquant) #value range: 0...75
plt.title("The_Quantization_Indices_of_the_Scalefactors")
plt.xlabel("The_Bark_Subbands")
plt.show()
plt.plot(yq) #value range: -4...4,
plt.title("The_Quantization_Indices_of_the_Subband_values")
plt.xlabel("The_MDCT_Subbands")
plt.show()
plt.plot(20*np.log10(np.abs(y[:,10])+1e-2))
plt.plot(20*np.log10(mT[:-1,10]+1e-2))
plt.title('Spectra_for_one_Block')
plt.plot(20*np.log10(np.abs(ydeq[:,10]-y[:,10])+1e-2))
plt.plot(20*np.log10(np.abs(ydeq[:,10])+1e-2))
plt.legend(('Magnitude_Original_Signal_Spectrum','Masking_Threshold',
'Magnitude_Spectrum_Reconstructed_Signal_Error', 'Magnitude_Spectrum_Reconstructed
    _Signal'))
```

143

```
plt.xlabel('MDCT_subband')
plt.ylabel("dB")
plt.show()

plt.specgram(x, NFFT=2048, Fs=2*np.pi)
plt.title('Spectrogram_of_Original_Signal')
plt.xlabel("Block_#")
plt.ylabel("Normalized_Frequency_(pi_is_Nyquist_freq.)")
plt.figure()
plt.specgram(xrek, NFFT=2048, Fs=2*np.pi)
plt.title('Spectrogram_of_Reconstructed_Signal')
plt.xlabel("Block_#")
plt.ylabel("Normalized_Frequency_(pi_is_Nyquist_freq.)")
```

We then let the entire program run with

```
python psyac_quantization.py
```

This plays back the sounds of the original and reconstructed signal, and produces Figs. 4.8, 4.9, 4.10, 4.11, and 4.12.

Observe that the spectrum and the spectrogram of the reconstructed signal look clearly different from the original, with many spectral components simply missing (which reduced the bit rate), but it sounds indeed very similar, which is the goal of the psychoacoustic model.

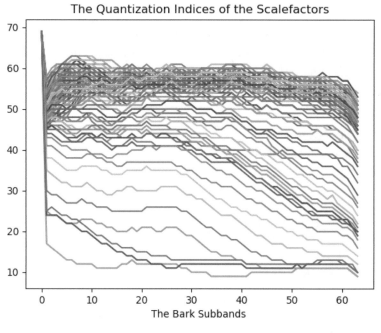

Figure 4.8: The quantization indices of the scalefactors over the Bark subbands. They are mostly between 10 and 70.

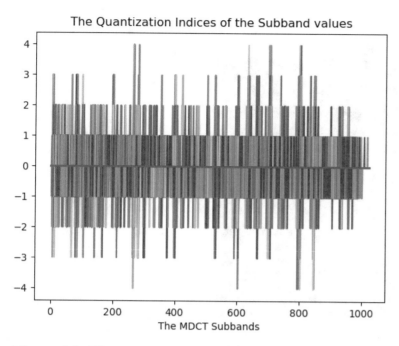

Figure 4.9: The quantization indices of the subband values over the MDCT subbands. They are mostly between −4 and 4.

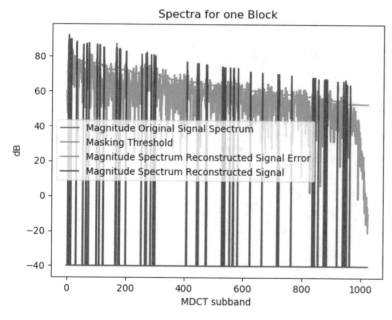

Figure 4.10: The spectra for a block. Observe the wide but still inaudible gaps in the reconstructed spectrum.

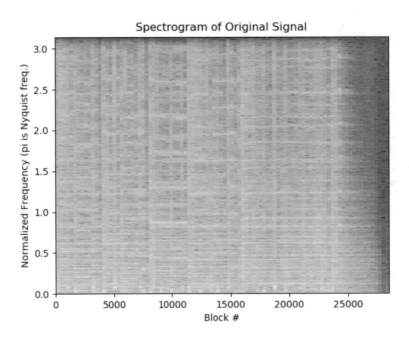

Figure 4.11: The spectrogram of the original.

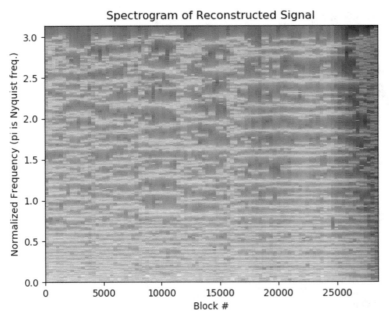

Figure 4.12: The spectrogram of the reconstructed signal. Observe the big holes in the spectrogram, which are still inaudible.

References

1. R.D. Patterson, I. Nimmo-Smith, J. Holdsworth, P. Rice, An efficient auditory filterbank based on the gammatone function, in *A Meeting of the IOC Speech Group on Auditory Modelling at RSRE* (1987)

2. S. Strah, A. Mertins, Analysis and design of gammatone signal models. Acoust. Soc. Am. **126**(5), 2379–2389 (2009)

3. E. Zwicker, H. Fastl, *Psychoacoustics, Facts and Models* (Springer, Berlin, 1990)

4. E. Zwicker, Subdivision of the audible frequency range into critical bands (Frequenzgruppen). J. Acoust. Soc. Am. **33**(2), 248 (1961)

5. H. Fletcher, Auditory patterns. Rev. Mod. Phys. **12**, 47–65 (1940)

6. J.O. Smith III, J.S. Abel, in *The Bark and ERB Bilinear Transforms*. Center for Computer Research in Music and Acoustics (CCRMA), Stanford University, 2007. Preprint of version accepted for publication in the IEEE Transactions on Speech and Audio Processing (1999)

7. M. Schroeder, Recognition of complex acoustic signals, in *Dahlem Konferenzen, Life Sciences Research Report 5*, ed. by T.H. Bullock (Abakon Verlag, Berlin, 1977), pp. 323–328

8. H. Hermansky, Perceptual linear predictive (PLP) analysis of speech. J. Acoust. Soc. Am. **87**(4), 1738–1752

9. T. Thiede, et al., PEAQ—the ITU standard for objective measurement of perceived audio quality. J. Audio Eng. Soc. **48**(1/2), 3–29 (2000)

10. E. Terhardt, Calculating virtual pitch. Hear. Res. **1**, 155–182 (1979)

11. U. Zoelzer, *Digital Audio Signal Processing* (Wiley, London, 2008)

12. A. Spanias, T. Painter, V. Atti, *Audio Signal Processing and Coding* (Wiley, London, 2005)

13. J. Johnston, Transform coding of audio signals using perceptual noise criteria. IEEE J. Sel. Areas in Commun. **6**(2), 314–323 (1988)

14. F. Baumgarte, C. Ferekidis, H. Fuchs, A nonlinear psychoacoustic model applied to the ISO MPEG layer 3 coder, in *99th AES Convention* (1995)

15. R.A. Lutfi, A power-law transformation predicting masking by sounds with complex spectra. J. Acoust. Soc. Am **77**(6), 2128–2136 (1985)

16. *MUSHRA*, https://en.wikipedia.org/wiki/MUSHRA. Accessed Jan 2019

17. *Recommendation BS.1116*, https://www.itu.int/rec/R-REC-BS.1116/en. Accessed Jan 2019

18. *PEAQ*, https://en.wikipedia.org/wiki/PEAQ. Accessed Jan 2019

19. R. Huber, B. Kollmeier, PEMO-Q—a new method for objective audio quality assessment using a model of auditory perception. EEE Trans. Audio Speech, Lang. Process. **14**, 1902–1911 (2006)

20. V. Emiya, E. Vincent, N. Harlander, V. Hohmann, Subjective and objective quality assessment of audio source separation. IEEE Trans. Audio Speech Lang. Process. **19**, 2046–2057 (2011)

21. MPEG, *Information Technology: Generic Coding of Moving Pictures and Associated Audio Information—Part 7: Advanced Audio Coding (AAC)* (ISO/IEC, 1997)

5 Entropy Coding

5.1 Introduction

After quantization of the subband values and the masking threshold or scalefactors we obtain the quantization indices. Finally we need to translate those indices into an Entropy code, which features variable length binary codewords for each index, for storing the information into a file. The goal of an Entropy coder is to minimize the total number of necessary bits for the file, to obtain the best compression. The definition of entropy: Take, for instance, the indices of our subband samples yq. They are integers i in a certain value range $i_{min}, \ldots, i_{max}$, for instance, in $-4, \ldots, 4$, as was the case in the previous section. That yq takes on the value of a specific i has a certain probability, which we call $p(i)$. This probability $p(i)$ can be computed as the number of times the index i appears in an audio signal, divided by the total number of samples we have, such that

$$\sum_{i=i_{min}}^{i_{max}} p(i) = 1$$

Then our entropy H is

$$H = -\sum_{i=i_{min}}^{i_{max}} p(i) \cdot \log_2(p(i))$$

Since we use the log of base 2, the result is interpreted as bits. We can then construct a binary Huffman coder, which encodes individual samples yq (sequence length $= 1$), such that average number of bits per sample R is similar to the entropy of yq,

$$H \leq R < H + 1$$

(see [1]).

5.2 Huffman Coding

An often used method for Entropy coding is Huffman coding, for instance, used in MPEG-AAC [2]. It constructs binary codewords whose length is proportional to the negative log of the probability of the index to encode. In this way the average number of bits per sample comes close to the entropy of the signal [1].

For Python we can install a Huffman coder library with

```
sudo pip3 install dahuffman
```

© Springer Nature Switzerland AG 2020
G. Schuller, *Filter Banks and Audio Coding*,
https://doi.org/10.1007/978-3-030-51249-1_5

We can then try it in a Python terminal with an example signal **a**. This library constructs the Huffman code directly from a sequence of indices, which means it internally computes the probability of each index, and uses it as the basis for the code. The encoding function object is "codec". The resulting code table can be printed with "print_code_table",

```
from dahuffman import HuffmanCodec
import numpy as np
a=np.array([-1,0,0,1,])
codec=HuffmanCodec.from_data(a)
table=codec.get_code_table()
codec.print_code_table()
bits  code       (value)  symbol
3   110        (    6)  _EOF
3   111        (    7)  -1
1   0          (    0)  0
2   10         (    2)  1
ac=codec.encode(a)
ac
b'\xe5'
```

Observe that there is also an "End of File" (EOF) symbol, which is important for a simple detection of the end of a sequence, and that we have indeed an unequal code length. More likely indices or symbols have shorter codewords. The actual Huffman encoding is done in the last line, with the binary Huffman code of our sequence stored in variable "ac", the resulting binary code (with the "b" in the beginning) in the last line. Observe how short it appears.

The corresponding decoder needs this code table (in the code the variable "table"), hence we need to store it for the decoding process, and the Huffman decoder needs to read it in.

```
dec=HuffmanCodec(code_table=table, check=False)
dec.decode(ac)
[-1, 0, 0, 1]
```

Observe that the decoder was indeed able to reconstruct the original sequence from the binary code.

Observe that this Huffman coder is made for one-dimensional array-like inputs and outputs, but our audio coder needs to encode two-dimensional arrays. Hence we need to "flatten" our 2-d arrays, and reshape them accordingly after decoding. The following code flattens our array of subband quantization indices in column order, such that the subband samples from one block stay together (the "−1" in reshape means "until the end"),

```
#Train with flattened quantized subband samples
yqflattened=np.reshape(yq,(1,−1),order='F')
yqflattened=yqflattened[0] #remove dimension 0
```

In the decoder, the 2-d shape needs to be reconstructed, which is done with the following code:

#reshape them back into a matrix with column length N:
yq=np.reshape(yqflattened, (N,−1),order='F')

In our Python audio coder, we use this program to first generate a Huffman codebook based on the file to encode, and then store the resulting codebook at the beginning of the compressed file, for the decoder. This has the advantage that it is adapted to the signal to encode, but it has the disadvantage that this file has to be fully known before storing or sending it, which makes streaming impossible. Also, it is not adapted to changing statistics inside the signal. For those reasons, MPEG-AAC uses fixed Huffman tables, which can be switched according to the local signal statistics [2].

5.3 Golomb–Rice Coding

The Golomb–Rice coder [3, 4] is a simple entropy coder for the case that the indices i to encode have a Laplacian probability distribution,

$$p(i) \propto \cdot e^{-|k \cdot i|}$$

for some constant k. In this case the entropy H or length of the optimum codeword for index i in bits is

$$H(i) \propto \log_2(p(i)) = k \cdot |i|$$

, hence the length of the codeword is proportional to the magnitude of our index i. To reflect the different growth of the length with different constants k, the Golomb–Rice coder uses 2 parts in its code. The first part encodes the rough size of the index i, which is obtained by dividing i by some "block size" M,

$$i_b = floor\left(\frac{i}{M}\right),$$

where "floor" is the rounding down operation. M can basically be any positive integer, but the code becomes simpler when it is a power of 2,

$$M = 2^b$$

The "block index" i_b is encoded using a "unary" code, which consists of a zero, followed by i_b ones. The second part encodes the "in-block index" $i - i_b \cdot M$, using a fixed length binary code, with sufficient length to encode all possible in-block indices.

The following shows a Python example for the case of $b = 1$,

#for installation: sudo pip install audio.coders
from audio.coders **import** rice
from bitstream **import** BitStream
import numpy as np

```
origs=np.arange(-2,6)
print("Original=_", origs)
#b: exponent of 2
ricecode=rice(b=1,signed=True)
riceencoded= BitStream(origs.astype(np.int32), ricecode)
print("rice_encoded=", riceencoded)

for index in origs:
    print("Index:_", index, "Rice_code:_", BitStream(index, ricecode))

ricedecoded=riceencoded.read(ricecode, 8)
print("rice_decoded=", ricedecoded)
```

It works with Python2.7. Running `python ricecodertest.py` produces the output,

```
('Original= ', array([-2, -1,  0,  1,  2,  3,  4,  5]))
('rice encoded=', 1010110000010001001100011001110)
('Index: ', -2, 'Rice code: ', 1010)
('Index: ', -1, 'Rice code: ', 110)
('Index: ', 0, 'Rice code: ', 000)
('Index: ', 1, 'Rice code: ', 010)
('Index: ', 2, 'Rice code: ', 0010)
('Index: ', 3, 'Rice code: ', 0110)
('Index: ', 4, 'Rice code: ', 00110)
('Index: ', 5, 'Rice code: ', 01110)
('rice decoded=', [-2, -1, 0, 1, 2, 3, 4, 5])
```

The Golomb–Rice coder is usually just for positive numbers. This example also shows how to encode signed integers, by just using the first code bit as a sign bit. The question how to choose the "block size" parameter b is answered in [5], as within the following range:

$$floor(\log_2(E(|i|))) \le b \le ceil(\log_2((E(|i|) + 1) \cdot 2/3)), \tag{5.1}$$

where "ceil" is the rounding up operation, and b should be limited to

$$0 \le b.$$

A good simple estimate of the Rice coefficient b is the average between the two boundaries, which we also chose for our example implementations,

$$b = 0.5 \cdot floor(\log_2(E(|i|))) + 0.5 \cdot ceil(\log_2((E(|i|) + 1) \cdot 2/3)) \tag{5.2}$$

where before the averaging it is ensured that the boundaries are non-negative.

References

1. R.G. Gallager, *Information Theory and Reliable Communication* (Wiley, 1968)

2. MPEG, *Information Technology: Generic Coding of Moving Pictures and Associated Audio Information—Part 7: Advanced Audio Coding (AAC)* (ISO/IEC, 1997)

3. D. Solomon, *Data Compression* (Springer, Berlin, 2000)

4. K. Sayood, *Introduction to Data Compression* (Morgan Kaufman, Los Altos, 2006)

5. *Rice Coding Coefficient*, https://ipnpr.jpl.nasa.gov/progress_report/42-159/159E.pdf. Accessed Sep 2018

6 The Python Perceptual Audio Coder

6.1 Introduction

Now we have all components we need to complete our audio coder, such that it can read in a mono or multi-channel audio signal, and write the compressed version into a file. The goal is to call it in a terminal window with the command

```
python3 audio_encoder.py audiofile.wav [quality]
```

where audiofile.wav is the audio file to compress, and "quality" is an optional quality argument. The default is quality $= 100$ as 100%. For higher values, the masking threshold is lowered accordingly, to reduce the quantization error and increase the quality, but also increase bit rate, and vice versa for lower quality values. For multi-channel audio signals, like stereo, it encodes the channels separately. This is simple, but not very efficient. More efficient approaches would be mid-side coding [1], or parametric spatial audio processing [2]. It writes the compressed signal to the file audiofile.acod.

For testing the coder, we can also record our own sound to a stereo audio file using our computer microphones. In Linux we use the following command for a stereo file with 32 kHz sampling rate and 16 signed bits per sample with Little Endian, for a duration of 5 s,

```
arecord -c 2 -r 32000 -f S16_LE -d 5 stereosound.wav
```
You can then listen to it using the command
```
aplay stereosound.wav.
```
We can also obtain free sound files from [3], for instance, "fantasy-orchestra.wav" for testing.

6.2 The Encoder

We begin our encoder program with importing the needed libraries. This includes the library "Pickle", which we will use to write a structured binary compressed file,

```
import sys
sys.path.append('./PythonPsychoacoustics')
from psyac_quantization import *
import numpy as np
import scipy.io.wavfile as wav
```

© Springer Nature Switzerland AG 2020
G. Schuller, *Filter Banks and Audio Coding*,
https://doi.org/10.1007/978-3-030-51249-1_6

```
import os
#sudo pip3 install dahuffman
from dahuffman import HuffmanCodec
if sys.version_info[0] < 3:
    # for Python 2
    import cPickle as pickle
else:
    # for Python 3
    import pickle
```

Next the program reads in the audio file name and the optional quantity argument,

```
audiofile=sys.argv[1]
print("audiofile=", audiofile)


if len(sys.argv) ==3:
    quality=float(sys.argv[2])
else:
    quality=100.0
```

then it reads in the audio file, and appends a dimension if it is mono, to always have a channels dimension,

```
try:
    channels=x.shape[1] #number of channels, 2 for stereo (2 columns in x)
except IndexError:
    channels=1 # 1 for mono
    x=np.expand_dims(x,axis=1) #add channels dimension 1
```

Next it defines the constants, MDCT sine window, and constructs the file name for the compressed file, Then it reads in the audio file, and appends a dimension if it is mono, to always have a channels dimension,

```
N=1024 #number of MDCT subbands
nfilts=64 #number of subbands in the bark domain
#Sine window:
fb=np.sin(np.pi/(2*N)*(np.arange(int(1.5*N))+0.5))


#Store in a pickle binary file:
#remove extension from file name:
name,ext=os.path.splitext(audiofile)
#new extension for compressed file:
encfile=name+'.acod'
print("Compressed_file:", encfile)
```

It then opens the compressed file, and uses "pickle" to write the sampling frequency and number of channels to encode into the beginning of the file,

```
with open(encfile, 'wb') as codedfile: #open compressed file
    pickle.dump(fs,codedfile) #write sampling rate
    pickle.dump(channels,codedfile) #write number of channels
```

Then we start a "for" loop over the channels of the audio signal and call our function for the MDCT, psycho-acoustics, and quantization for each channel,

```
for chan in range(channels): #loop over channels:
    print("channel_", chan)
    #Compute quantized masking threshold in the Bark domain and quantized subbands
    yq, y, mTbarkquant=MDCT_psayac_quant_enc(x[:,chan],fs,fb,N, nfilts,quality=quality)
```

After that we can construct the Huffman code tables, and Huffman encode the quantized masking threshold in the Bark domain (scalefactors) "mTbarkquant" and also the quantized subband values yq.

```
    print("Huffman_Coding")
    #Train Huffman coder for quantized masking threshold in the Bark domain (scalefactors),
    #with flattening the masking threshold array in column (subband) order:
    mTbarkquantflattened=np.reshape(mTbarkquant, (1,−1),order='F')
    mTbarkquantflattened=mTbarkquantflattened[0] #remove dimension 0
    codecmTbarkquant=HuffmanCodec.from_data(mTbarkquantflattened)
    #Huffman table for it:
    tablemTbarkquant=codecmTbarkquant.get_code_table()
    #Huffman encoded:
    mTbarkquantc=codecmTbarkquant.encode(mTbarkquantflattened)

    #Compute Huffman coder for the quantized subband values:
    #Train with flattened quantized subband samples
    yqflattened=np.reshape(yq,(1,−1),order='F')
    yqflattened=yqflattened[0] #remove dimension 0
    codecyq=HuffmanCodec.from_data(yqflattened)
    #Huffman table for it:
    tableyq=codecyq.get_code_table()
    #Huffman encoded:
    yqc=codecyq.encode(yqflattened)
```

Finally we can write such encoded data to our pickle file,

```
    pickle.dump(tablemTbarkquant ,codedfile) #scalefactor Huffman table
    pickle.dump(tableyq ,codedfile) #subband sample Huffman table
    pickle.dump(mTbarkquantc ,codedfile) #Huffman coded scalefactors
    pickle.dump(yqc ,codedfile) #Huffman coded subband samples
```

Now we are done and can test our program on our test audio file "sc03_16m.wav" (with 16 kHz sampling rate), or any other audio .wav file,

```
python3 audio_encoder.py sc03_16m.wav
```

It writes the compressed file "sc03_16m.acod". We can now take a look at the file sizes. The original "sc03_16m.wav" had a size of 357,756 bytes, the compressed file "sc03_16m.acod" has a size of 37,530 bytes. This means we obtained a compression factor of 357,756/37,530 = 9.53. Since our signal contains 178,856 samples, this is 37,530 * 8/178,856 = 1.68 bits/sample on average.

6.3 The Decoder

Our audio decoder should decode our encoded audio file back into a .wav file. We call it in a terminal window with the command

```
python3 audio_decoder.py audiofile.acod
```

It writes the decoded audio signal in file audiofilerek.wav

We again start with loading the required libraries, read the name of the encoded audio file, set the constants, and produce the name for the reconstructed audio file,

```
import sys
sys.path.append('./PythonPsychoacoustics')
from psyac_quantization import *
import numpy as np
import scipy.io.wavfile as wav
import os
from dahuffman import HuffmanCodec
if sys.version_info[0] < 3:
    # for Python 2
    import cPickle as pickle
else:
    # for Python 3
    import pickle

if len(sys.argv) < 2:
    print("Usage:_python3_audio_decoder_audiofile.acod")

encfile=sys.argv[1]
print("encoded_file=",encfile)

N=1024 #number of MDCT subbands
nfilts=64 #number of subbands in the bark domain
#Sine window:
fb=np.sin(np.pi/(2*N)*(np.arange(int(1.5*N))+0.5))

#Open the pickle binary file:
#remove extension from file name:
name,ext=os.path.splitext(encfile)
#new name end extension for decoded file:
decfile=name+'rek.wav'
print("Decoded_file:", decfile)
```

The program opens the encoded file, reads in the sampling frequency and the number of channels, and starts the "for" loop over the number of channels. In the "for" loop, for each channel the Huffman tables are read in, and also the Huffman coded scalefactors and subband samples. Then the Huffman decoder is applied to the scalefactors, which are then re-shaped to the original array shape. The same is then done with the subband samples.

with **open**(encfile, 'rb') as codedfile: #*open compressed file*
    fs=pickle.load(codedfile)
    channels=pickle.load(codedfile)
    print("fs=", fs, "channels=", channels,)

    for chan **in range**(channels): #*loop over channels:*
        print("channel_", chan)
        tablemTbarkquant=pickle.load(codedfile) #*scalefactor Huffman table*
        tableyq=pickle.load(codedfile) #*subband sample Huffman table*
        mTbarkquantc=pickle.load(codedfile) #*Huffman coded scalefactors*
        yqc=pickle.load(codedfile) #*Huffman coded subband samples*

        #*Huffman decoder for the scalefactors:*
        codecmTbarkquant=HuffmanCodec(code_table=tablemTbarkquant, check=False)
        #*Huffman decoded scalefactors:*
        mTbarkquantflattened=codecmTbarkquant.decode(mTbarkquantc)
        #*reshape them back into a matrix with column length nfilts:*
        mTbarkquant=np.reshape(mTbarkquantflattened, (nfilts,−1),order='F')

        #*Huffman decoder for the subband samples:*
        codecyq=HuffmanCodec(code_table=tableyq, check=False)
        #*Huffman decode the subband samples:*
        yqflattened=codecyq.decode(yqc)
        #*reshape them back into a matrix with column length N:*
        yq=np.reshape(yqflattened, (N,−1),order='F')

Finally these are input to our function `MDCTsyn_dequant_dec`, which contains the dequantization and the synthesis MDCT, and outputs the reconstructed audio channel. In case of a multi-channel signal, the channels are stacked such that each column is one channel or the audio signal. Then the "clip" function is applied, to ensure that we produce no overflow in the sound file (for signed 16 bit integers for the samples), and the resulting sound array is stored in our reconstructed sound file, in 16 bit integer sample format.

        xrek, mT, ydeq = MDCTsyn_dequant_dec(yq, mTbarkquant, fs, fb, N, nfilts)
        if chan==0:
            x=xrek
        else:
            x=np.vstack((x,xrek))
x=np.clip(x.T,−2**15,2**15−1)
#*Write decoded signal to wav file:*
wav.write(decfile,fs,np.int16(x))

We can now try it with decoding our previously encoded file with the command,

```
python3 audio_decoder.py sc03_16m.acod
```

and under Linux play it with

```
play sc03_16mrek.wav
```

and compare it to the original. We should now hear the decoded sound with only barely audible slight distortions, or just noticeable differences, which means our distortions are just at the masking threshold. Also note that the decoder is faster than the encoder, because it does not need to compute the psycho-acoustic model. Also note that our simple audio coder has no window- or subband switching, it has a constant number of subbands. Hence there will be audible pre-echo artifacts for attacks like in castanet or drum signals, or reverberation artifacts for speech at $N = 1024$ subbands. For experimenting, we can reduce this number, for instance, to $N = 128$, which removes the pre-echos, but observe that also increases the bit rate.

Now we could try several different audio files, stereo or mono (in the stereo case, each stereo channel is just encoded separately), for instance, a tonal signal, a percussive signal, and a speech signal. For each find the quality setting at which the audible distortions just disappear, and observe the bit rates (in bits per sample). For percussive sounds and speech also try lower number of subbands, like $N = 256$ or $N = 128$, on line 48 of `audio_encoder.py` and line 27 of `audio_decoder.py`, and hear if the pre-echos disappear, or for more tonal signals increase it to $N = 2048$, and see if the bit rate is reduced. We can also try to reduce the number of Bark subbands, for instance, to `nfilts=48`, on line 49 in the encoder and line 28 in the decoder, or vary the `alpha` factor, for instance, to `alpha=0.7` on line 28 of `psyac_quantization.py`, and hear if and what effect it has on audible distortions.

References

1. Y. Huang, J. Benesty (eds.), *Audio Signal Processing for Next-Generation Multimedia Communication Systems* (Kluwer Academic Publishers, Dordrecht, 2004)

2. J. Breebart, C. Faller, *Spatial Audio Processing: MPEG Surround and Other Applications* (Wiley, London, 2007)

3. *Freesound*, https://freesound.org/. Accessed Jan 2019

7 Predictive Lossless Audio Coding

7.1 Introduction

Unlike perceptual coding, where the decoded signal should sound to the human ear like the original, the goal of lossless coding is to reconstruct the original audio samples exactly. Here we assume that the original audio signal consists of integer valued samples, but the methods can equally be applied to float valued samples, since they also only have a finite set of values. The most commonly used approach for lossless audio coding is predictive coding (Sect. 3.3.5), which, for instance, the "Free Lossless Audio Coder" (FLAC) [1] is using, see also [2]. The following describes an example implementation of a lossless predictive coder based on the LMS principle in Python.

7.2 The Predictive Lossless Encoder

In this section we take the lossless predictor of Sect. 3.3.6 and combine it with the Rice coder of Sects. 5.3 and 8.3 to obtain a complete predictive lossless audio coder. Observe that the Rice coding library needs Python2. We start with the NLMS predictor of Sect. 3.3.6 and write it as a function,

```
def nlmslosslesspredenc(x,L,h):
    #Computes the NLMS lossless predictor
    #arguments: x: input signal (mono)
    #L: Predictor lenght
    #h: starting values for the L predictor coefficients
    #returns: e, the prediction error
    x=np.hstack((np.zeros(L),x)); #make same starting conditions as decoder
    e=np.zeros(len(x));
    for n in range(L, len(x)):
        #prediction error and filter, using the vector of reconstructed samples,
        #predicted value from past reconstructed values, since it is lossless, xrek=x:
        xrekvec=x[n−L+np.arange(L)]
        P=np.dot(np.flipud(xrekvec), h);
        #quantize and de−quantize by rounding to the nearest integer:
        P=round(P)
        #prediction error:
        e[n]=x[n]−P
        #NLMS update:
        h = h + 1.0* e[n]*np.flipud(xrekvec)/(0.1+np.dot(xrekvec,xrekvec))
```

© Springer Nature Switzerland AG 2020
G. Schuller, *Filter Banks and Audio Coding*,
https://doi.org/10.1007/978-3-030-51249-1_7

```
    #if n%100==0: print("h=", h, "P=", P, "e[n]=", e[n], "xrekvec=", xrekvec)
  return e
```

In the main section of the program we start with reading the audio ".wav" file name from the argument, read it in, make a two-dimensional array out of the audio samples even in the mono case, and construct the file name of the resulting encoded file.

```
if __name__ == '__main__':
  if (len(sys.argv) <2):
    print('\033[93m'+"Need_audio_file_as_argument!"+'\033[0m') #warning in yellow font

  audiofile=sys.argv[1]
  print("audiofile=", audiofile)

  fs, x= wav.read(audiofile)
  x=x*1.0 #make it float to avoid overflow!
  print("Sampling_Frequency_in_Hz=", fs, "max(x)=", np.max(x))

  try:
    channels=x.shape[1] #number of channels, needs to be 2 for stereo (2 columns in x)
  except IndexError:
    channels=1 # 1 for mono, make x also 2-dimensional (chan is last dim):
    x=np.expand_dims(x,axis=-1)

  print("channels=", channels, "x.shape=", x.shape)

  N=int(fs*20e-3) #fs*20ms=640, number of samples per rice coder block
  L=10 #Predictor order

  #Store in a pickle binary file:
  #remove extension from file name:
  name,ext=os.path.splitext(audiofile)
  #new extension for compressed file:
  encfile=name+'.lacodpred'
  print("Compressed_file:", encfile)
```

Here we also set the length of our predictor to $L = 10$, and the length per Rice coding block to $20\,\mathrm{ms}$, which is 640 samples at $32\,\mathrm{kHz}$ sampling rate. $20\,\mathrm{ms}$ is a common block length for speech processing, because it is assumed that a speech signal is more or less stationary within such a block. For each such block we later compute the best Rice coefficient according to Eq. (5.2) and send it as side information to the decoder. The encoded file is structured by the library "pickle". We open the encoded file and in the beginning we write the sampling rate and the number of channels using "pickle.dump".

```
  with open(encfile, 'wb') as codedfile: #open compressed file
    pickle.dump(fs, codedfile, protocol=-1) #write sampling rate
    pickle.dump(channels, codedfile, protocol=-1) #write number of channels
```

Then we loop over each audio channel, compute the prediction error, reshape it such that each Rice coding block appears in a column of array "prederror", compute the Rice coefficients according to Eq. (5.2) for each column, and store them as 8 bit integers in the encoded file, again using "pickle.dump".

```
for chan in range(channels): #loop over channels:
    print("channel_", chan)
    print("NLMS_prediction:")
    h=np.zeros(L)
    e=nlmslosslesspredenc(x[:,chan],L,h) #compute the NLMS predicton error
    print("len(e)", len(e))
    numblocks=len(e)//N
    prederror=np.reshape(e[:numblocks*N], (N,numblocks), order='F')
    print("numblocks=", numblocks)
    if chan==0: pickle.dump(numblocks, codedfile, protocol=-1) #write number of
      blocks
    print("Rice_Coding:")
    #Suitable Rice coding coefficient estimation for the blocks:
    #https://ipnpr.jpl.nasa.gov/progress_report/42-159/159E.pdf
    meanabs=np.mean(np.abs(prederror),axis=0)
    ricecoefff=np.clip(np.floor(np.log2(meanabs)),0,None)
    ricecoeffc=np.clip(np.ceil(np.log2((meanabs+1)*2/3)),0,None)
    ricecoeff=np.round((ricecoeffc+ricecoefff)/2.0).astype(np.int8) #integer, 8bit
    #print("ricecoeff=", ricecoeff)
    s=struct.pack('b'*int(len(ricecoeff)),*ricecoeff)
```

Next the program adds 4 zeros to each block of prediction errors to possibly complete Rice bytes, as a precaution because we can only write full bytes to file, and we will lose partly filled bytes. Then it loops over the Rice coding blocks, Rice-encodes each of them with their own Rice coefficient, and writes them to the encoded file,

```
prederror=np.concatenate((prederror, np.zeros((4,numblocks))), axis=0)
  #add 4 zeros
#to each block of prediction errors to possibly complete rice bytes.
for k in range(numblocks): #loop across blocks:
    if (k%100==0): print("block:",k)
    #Rice coding with m=2**b
    signedrice=rice(b=ricecoeff[k],signed=True)
    prederrorrice= BitStream(prederror[:,k].astype(np.int32), signedrice)
    #see: http://boisgera.github.io/bitstream/
    #Turn bitstream format into sequence of bytes:
    prederrors=prederrorrice.read(bytes, np.floor(len(prederrorrice)/8.0))
```

To execute our predictive lossless coder on an example audio file "fspeech.wav" (with 16 bit per sample and 32 kHz sampling rate), we execute in a terminal shell,

`python lossless_predictive_audio_encoder.py fspeech.wav`

It produced the file "fspeech.lacodpred". We can observe that its size is a factor 0.569 of the size of the original file. Instead of "fspeech.wav", we could also take a file from [3].

7.3 The Predictive Lossless Decoder

The decoder uses the lossless predictor from Sect. 3.3.6 for reconstruction as a function

```
def nlmslosslesspreddec(e,L,h):
    #Computes the NLMS lossless predictor
    #arguments: x: input signal (mono)
    #L: Predictor lenght
    #h: starting values for the L predictor coefficients
    #returns: e, the prediction error

    xrek=np.zeros(len(e))
    for n in range(L, len(e)):
        #prediction error and filter, using the vector of reconstructed samples,
        #predicted value from past reconstructed values, since it is lossless, xrek=x:
        xrekvec=xrek[n−L+np.arange(L)]
        P=np.dot(np.flipud(xrekvec), h);
        #quantize and de−quantize by rounding to the nearest integer:
        P=round(P)
        #reconstructed value from prediction error:
        xrek[n] = e[n] + P;
        #NLMS update:
        h = h + 1.0* e[n]*np.flipud(xrekvec)/(0.1+np.dot(xrekvec,xrekvec))
    return np.hstack((xrek[L:],np.zeros(L))) #remove leading zeros to avoid their delay
```

The main part reads the compressed file name from the argument of the program, opens it, and creates the reconstructed file name,

```
if __name__ == '__main__':
    if (len(sys.argv) <2):
        print('\033[93m'+"Need_*.lacodpred_encoded_audio_file__as_argument!"+'\033[0m')
            #warning, yellow font
        sys.exit();
    encaudiofile=sys.argv[1]
    print("encoded_audiofile=", encaudiofile)
    #Store decoded audio in file:
    #remove extension from file name:
    name,ext=os.path.splitext(encaudiofile)
    print("Extension=", ext)
    if ext != '.lacodpred':
        print('\033[93m'+"Need_*.lacodpred_encoded_audio_file!"+'\033[0m')
            #warning, yellow font
        sys.exit();
    #new name end extension for decoded file:
    decfile=name+'larek.wav'
```

Then it opens the encoded file, reads in the sampling rate, computes the block size from it, reads in the number of channels and the number of blocks in each channel, and then loops over the channels and blocks for reconstruction,

L=10 *#Predictor order*

```
with open(encaudiofile, 'rb') as codedfile: #open compressed file
    fs=pickle.load(codedfile)
    N=int(fs*20e-3) #fs*20ms=640, number of samples per rice coder block
    channels=pickle.load(codedfile)
    print("fs=", fs, "channels=", channels, )
    numblocks=pickle.load(codedfile)
    #N-=4 #last encoded samples might be missing from rounding to bytes
    print("numblocks=", numblocks)
    xrek=np.zeros((numblocks*N, channels))
    for chan in range(channels): #loop over channels:
        print("channel_", chan)
        ricecoeffcomp=pickle.load(codedfile);
        ricecoeff =struct.unpack( 'B' * len(ricecoeffcomp), ricecoeffcomp);
        print("len(ricecoeff)=", len(ricecoeff))
        prederrordec=np.zeros(N*numblocks)
        for k in range(numblocks): #loop across blocks:
            if (k%100==0): print("Block_number:",k)
            prederrors=pickle.load(codedfile) #Rice coded block samples
            #m=2**b
            signedrice=rice(b=ricecoeff[k],signed=True)
            prederrorrice = BitStream();
            prederrorrice.write(prederrors)
            prederrordec[k*N:(k+1)*N]=prederrorrice.read(signedrice, N);
        print("NLMS_prediction:");
        h=np.zeros(L)
        prederrordec=prederrordec*1.0 #convert to float to avoid overflow
        print("len(prederrordec)=", len(prederrordec))
        xrek[:len(prederrordec),chan]=nlmslosslespreddec(prederrordec,L,h)
```

Now we can execute the decoder with the command
`python lossless_predictive_audio_decoder.py fspeech.lacodpred`
It produces the file `fspeechlarek.wav`, which is identical to the original, but a few samples shorter, because the encoder only processes full Rice coding blocks (the line `numblocks=len(e)//N`). Expect a compression factor of about 2, depending on the signal. The better a signal is to predict, the higher the compression factor we obtain.

References

1. *FLAC*, https://en.wikipedia.org/wiki/FLAC. Accessed Sep 2018

2. A. Spanias, T. Painter, V. Atti, *Audio Signal Processing and Coding* (Wiley, London, 2005)

3. *Freesound*, https://freesound.org/. Accessed Jan 2019

8 Scalable Lossless Audio Coding

8.1 Introduction

Lossless audio coding has as a goal to reconstruct the exact same audio samples after decoding. Hence we cannot use psycho-acoustic models, but have to rely on redundancy reduction, the correlations between samples. This has the advantage that the encoder structure is relatively simple, but the disadvantage that the resulting bit rate or size of the compressed file depends wholly on the statistics of the audio source file, and can even be bigger than the source file, for instance, if random samples are used as source file. Often there are limitations in the file size, or in the bit rate, and hence it is desirable to have a scalable lossless coder, which can be lossless if bit rate permits, and nearly lossless (meaning with only small errors in the reconstructed samples) otherwise. This is why the MPEG Scalable Lossless (SLS) audio coder [1] was developed and standardized.

8.2 The Integer-to-Integer MDCT

Instead of predictive coding, MPEG-SLS is based on integer-to-integer transform coding. Its analysis filter bank produces integer valued subband values if the audio source consists of integer valued samples. This is done in such a way that its synthesis filter bank reconstructs the original integer samples of the audio source. This can then be used as an extension layer to the MPEG-AAC as a core coder, if the integer-to-integer transform is an integer-to-integer MDCT (IntMDCT) filter bank with the same window function and number of subbands as the MPEG-AAC core coder. Then we can construct a lossless enhancement layer by taking the difference between the IntMDCT subband values and the MDCT subband values (after de-quantization). This gives us 2 levels of encoding: a lossy level from the MPEG-AAC, and the lossless enhancement level. In the next step, we can introduce more bit rate steps for the enhancement layer, to obtain more levels between the MPEG-AAC and lossless coding. This could be used, for instance, to obtain more robustness against quality degradation from several encoding/decoding steps (tandem coding), as can happen with several processing steps in studios.

In the following the IntMDCT will be described. The first problem to solve is how to obtain an integer-to-integer transform, which is invertible for integer valued inputs. We cannot simply use rounding after the analysis and synthesis filter banks, because that could lead to different integer values after the synthesis filter bank. An elegant way to obtain an invertible integer-to-integer transform is the so-called lifting step [2, 3]. As an

© Springer Nature Switzerland AG 2020
G. Schuller, *Filter Banks and Audio Coding*,
https://doi.org/10.1007/978-3-030-51249-1_8

example, take a 2-subband transform, with blocks of 2 samples, $[x_0(m), x_1(m)]$. We can then apply a lifting step with some real valued factor a as a transform matrix as

$$[y_0(m), y_1(m)] = [x_0(m), x_1(m)] \cdot \begin{bmatrix} 1 & 0 \\ a & 1 \end{bmatrix}$$

The inverse lifting step is using the inverse matrix,

$$[x_0(m), x_1(m)] = [y_0(m), y_1(m)] \cdot \begin{bmatrix} 1 & 0 \\ -a & 1 \end{bmatrix}$$

Observe that the inverse matrix or lifting step looks almost the same as the forward step, except that the factor a now has a negative sign. This means we can easily include rounding with the factor, and as long as we use the exact same rounding in the forward and inverse step, we have perfect reconstruction. Writing it as individual equations, now including the rounding, the forward lifting step becomes

$$y_0(m) = x_0(m) + round(x_1(m) \cdot a)$$
$$y_1(m) = x_1(m)$$

and the inverse lifting step is

$$x_0(m) = y_0(m) - round(y_1(m) \cdot a)$$
$$x_0(m) = y_1(m).$$

Since we assume that we have integer valued input signals $x_0(m), x_1(m)$, we know that $round(x_i(m)) = x_i(m)$, and we can apply rounding after the entire matrix multiplication. Hence the forward lifting step with rounding becomes

$$[y_0(m), y_1(m)] = round\left([x_0(m), x_1(m)] \cdot \begin{bmatrix} 1 & 0 \\ a & 1 \end{bmatrix}\right)$$

and the inverse lifting step

$$[x_0(m), x_1(m)] = round\left([y_0(m), y_1(m)] \cdot \begin{bmatrix} 1 & 0 \\ -a & 1 \end{bmatrix}\right)$$

This means, we now have a simple, elementary, invertible integer-to integer transform with perfect reconstruction. If we now can decompose a matrix into a product of such lifting steps, we can turn it into an integer-to-integer transform [4, 5]

Now we take a look at our F matrix of the MDCT. It has 2×2 submatrices of the form of Eq. (1.55),

$$\begin{bmatrix} h(2N-1-n) & h(N-1-n) \\ h(N+n) & -h(n) \end{bmatrix}$$

and assume we have a sine window,

$$h(n) = \sin\left(\frac{\pi}{2N}(n+0.5)\right)$$

then the matrix becomes

$$
\begin{bmatrix} \sin\left(\frac{\pi}{2N}(2N-1-n+0.5)\right) & \sin\left(\frac{\pi}{2N}(N-1-n+0.5)\right) \\ \sin\left(\frac{\pi}{2N}(N+n+0.5)\right) & -\sin\left(\frac{\pi}{2N}(n+0.5)\right) \end{bmatrix}
$$

$$
= \begin{bmatrix} \sin\left(\frac{\pi}{2N}(n+0.5)\right) & \cos\left(\frac{\pi}{2N}(n+0.5)\right) \\ \cos\left(\frac{\pi}{2N}(n+0.5)\right) & -\sin\left(\frac{\pi}{2N}(n+0.5)\right) \end{bmatrix}
$$

using trigonometric identities ($\sin(-x) = -\sin(x)$, $\sin(\pi/2 + x) = \cos(x)$). With $\alpha := \frac{\pi}{2N}(n+0.5)$ this has the general form

$$
\begin{bmatrix} \sin(\alpha) & \cos(\alpha) \\ \cos(\alpha) & -\sin(\alpha) \end{bmatrix}.
$$

This is a rotation matrix [6] with the rows exchanged or flipped. This flipped rotation matrix can now be decomposed into a product of lifting steps, with their columns exchanged or flipped. In this way they have the basic shape of the Zero-Delay matrices (1.64), in the following way,

$$
\begin{bmatrix} \sin(\alpha) & \cos(\alpha) \\ \cos(\alpha) & -\sin(\alpha) \end{bmatrix}
$$

$$
= \begin{bmatrix} 0 & 1 \\ 1 & \frac{\cos(\alpha)-1}{\sin(\alpha)} \end{bmatrix} \cdot \begin{bmatrix} 0 & 1 \\ 1 & \sin(\alpha) \end{bmatrix} \cdot \begin{bmatrix} 0 & 1 \\ 1 & \frac{\cos(\alpha)-1}{\sin(\alpha)} \end{bmatrix}
$$

This can also be verified, for instance, using python.sympy, as shown in the main section of function `LiftingFmat.py`,

```
#With sympy:
import sympy
alpha=sympy.symbols('alpha')
F0=sympy.Matrix([[0,1],[1,(sympy.cos(alpha)-1)/sympy.sin(alpha)]])
L0=sympy.Matrix([[0,1],[1,sympy.sin(alpha)]])
L1=F0
print("Using sympy: flipped lifting steps:")
#print(sympy.latex(F0)) #for LaTeX
sympy.pprint(F0)
sympy.pprint(L0)
sympy.pprint(L1)
print("Their product is a flipped rotation matrix:")
sympy.pprint(sympy.simplify(F0*L0*L1))
```

When we increase the number of subbands N, we need to make sure that we obtain our diamond like matrix shape. For that we create the first matrix $F0$ in the product such that it has this desired shape. The following matrices $L0$ and $L1$ with their bi-diagonal shapes then maintain this diamond shape for the product matrix. This can be seen from the result of the following code piece from the main (testing) section in `LiftingMat.py`,

```
N=4 #number of subbands
fb=np.sin(np.pi/(2*N)*(np.arange(int(1.5*N))+0.5))

F0,L0,L1=LiftingFmat(fb)
print("Lifting_matrices_for_N=4:")
print("F0=\n", F0)
print("L0=\n", L0)
print("L1=\n", L1)

Fmat=np.dot(F0,L0)
Fmat=np.dot(Fmat,L1)
print("N=4:_Fmat_from_lifting=\n", Fmat)

#comparison to the usual symmetric F matrix:
from MDCTfb import symFmatrix
Fa=symFmatrix(fb)
print("for_comparison:_usual_Fa=_\n", Fa[:,:,0])
```

Both the previous and this part of the main section are executed if we execute from a terminal shell,
`python3 LiftingMat.py`.
The result of the latter part is

```
Lifting matrices for N=4:
F0=
[[ 0.          0.          1.          0.        ]
 [ 0.          0.          0.          1.        ]
 [ 1.          0.          0.         -0.30334668]
 [ 0.          1.         -0.0984914   0.        ]]
L0=
[[ 0.          0.          0.          1.        ]
 [ 0.          0.          1.          0.        ]
 [ 0.          1.          0.19509032  0.        ]
 [ 1.          0.          0.          0.55557023]]
L1=
[[ 0.          0.          0.          1.        ]
 [ 0.          0.          1.          0.        ]
 [ 0.          1.         -0.0984914   0.        ]
 [ 1.          0.          0.         -0.30334668]]
N=4: Fmat from lifting=
[[ 0.          0.19509032  0.98078528  0.        ]
 [ 0.55557023  0.          0.          0.83146961]
 [ 0.83146961  0.          0.         -0.55557023]
 [ 0.          0.98078528 -0.19509032  0.        ]]
for comparison: usual Fa=
[[ 0.          0.19509032  0.98078528  0.        ]
```

```
[ 0.55557023  0.          0.          0.83146961]
[ 0.83146961  0.         -0.         -0.55557023]
[ 0.          0.98078528 -0.19509032 -0.          ]]
```

Since the last 2 matrices are identical, it confirms that this decomposition into our product indeed works. It uses the following function `LiftingFmat` to construct the lifting matrices. In the beginning it applies the arcsin function to the coefficients of the sine window to obtain the needed rotation angles,

def LiftingFmat(fb):
    #produces the 3 lifting matrices F0, L0, L1,
    #whose product results in the F folding matrix for the IntMDCT.
    #Usage: F0,L0,L1=LiftingFmat(fb)
    #Argument fb: The 1.5N coefficients for the normal F matrix of the MDCT

    N=**int**(**len**(fb)/1.5)
    alpha=np.arcsin(fb[:N//2])
    #print("alpha=", alpha)
    #Lifting1:
    F0=np.zeros((N,N))
    #third quadrant anti−diagonal:
    F0[**int**(N/2):N,0:**int**(N/2)] =(np.eye(**int**(N/2)))
    #2nd quadrant anti−diagonal:
    F0[0:**int**(N/2),**int**(N/2):N] =(np.eye(**int**(N/2)))
    #4th quadrant:
    F0[**int**(N/2):N,**int**(N/2):N]=np.flipud(np.diag((np.cos(alpha)−1)/np.sin(alpha),k=0))

    #Lifting2:
    L0=np.zeros((N,N))
    #2nd quadrant anti−diagonal:
    L0[0:**int**(N/2),**int**(N/2):N] =np.fliplr(np.eye(**int**(N/2)))
    #3rd quadrant anti−diagonal:
    L0[**int**(N/2):N,0:**int**(N/2)] =np.fliplr(np.eye(**int**(N/2)))
    #4th quadrant:
    L0[**int**(N/2):N,**int**(N/2):N]=(np.diag(np.sin(alpha),k=0))

    #Lifting3:
    L1=np.zeros((N,N))
    #third quadrant anti−diagonal:
    L1[**int**(N/2):N,0:**int**(N/2)] =np.fliplr(np.eye(**int**(N/2)))
    #2nd quadrant anti−diagonal:
    L1[0:**int**(N/2),**int**(N/2):N] =np.fliplr(np.eye(**int**(N/2)))
    #4th quadrant:
    L1[**int**(N/2):N,**int**(N/2):N]=np.diag((np.cos(alpha)−1)/np.sin(alpha),k=0)
    return F0, L0, L1

Now we have the $\mathbf{F}$ folding matrix in Integer-to-Integer format. Next we need to implement the second part of our MDCT, the DCT4 transform matrix $\mathbf{T}$ of size $N \times N$, in Integer-to-Integer format. For that we apply a trick. Instead of applying it to a mono audio channel, we apply it to a stereo channel pair together [7]. Assume we have those 2 audio channels, after processing them through our Integer-to-Integer $\mathbf{F}$ folding matrix. Then we have a pair of signal arrays (which contain the signal blocks in its rows), one for the left and one for the right audio channel, which we call $\hat{\mathbf{x}}\mathbf{L}$ and $\hat{\mathbf{x}}\mathbf{R}$, and which we can arrange in a bigger array,

$$[\hat{\mathbf{x}}\mathbf{L}, \hat{\mathbf{x}}\mathbf{R}].$$

Applying the transform matrix $\mathbf{T}$ to this stereo pair means to multiply it with a larger matrix where the 2 transform matrices appear on the diagonal, to result in the MDCT subband arrays $\mathbf{yL}$ and $\mathbf{yR}$,

$$[\mathbf{yL}, \mathbf{yR}] = [\hat{\mathbf{x}}\mathbf{L}, \hat{\mathbf{x}}\mathbf{R}] \cdot \begin{bmatrix} \mathbf{T} & \mathbf{0} \\ \mathbf{0} & \mathbf{T} \end{bmatrix}$$

We have the special case that the DCT4 is orthonormal and symmetric, it means that its inverse is identical to the forward transform,

$$\mathbf{T}^{-1} = \mathbf{T}.$$

For the result we can also switch the left and right audio channels, which means we exchange the two rows for the transform matrix (and with it the inputs), and change the sign of the resulting left subbands. Then we obtain

$$[-\mathbf{yL}, \mathbf{yR},] = [\hat{\mathbf{x}}\mathbf{R}, \hat{\mathbf{x}}\mathbf{L}] \cdot \begin{bmatrix} \mathbf{0} & \mathbf{T} \\ -\mathbf{T}^{-1} & \mathbf{0} \end{bmatrix}$$

It turns out that we can again also decompose this larger transform matrix into a product of 3 Lifting steps,

$$\begin{bmatrix} \mathbf{0} & \mathbf{T} \\ -\mathbf{T}^{-1} & \mathbf{0} \end{bmatrix} = \begin{bmatrix} \mathbf{1} & \mathbf{T} \\ \mathbf{0} & \mathbf{1} \end{bmatrix} \cdot \begin{bmatrix} \mathbf{1} & \mathbf{0} \\ -\mathbf{T}^{-1} & \mathbf{1} \end{bmatrix} \begin{bmatrix} \mathbf{1} & \mathbf{T} \\ \mathbf{0} & \mathbf{1} \end{bmatrix},$$

where $\mathbf{0}$ and $\mathbf{1}$ are $N \times N$ zero- and identity matrices. Observe that this is a "multi-dimensional Lifting", since the entries of the matrices are again matrices of dimension $N \times N$. By using this decomposition we obtain

$$[-\mathbf{yL}, \mathbf{yR}] = [\hat{\mathbf{x}}\mathbf{R}, \hat{\mathbf{x}}\mathbf{L}] \cdot \begin{bmatrix} \mathbf{1} & \mathbf{T} \\ \mathbf{0} & \mathbf{1} \end{bmatrix} \cdot \begin{bmatrix} \mathbf{1} & \mathbf{0} \\ -\mathbf{T}^{-1} & \mathbf{1} \end{bmatrix} \begin{bmatrix} \mathbf{1} & \mathbf{T} \\ \mathbf{0} & \mathbf{1} \end{bmatrix}$$

We can rewrite this as individual equations for the 2 audio channels, now including rounding,

$$\mathbf{y0} := \mathbf{\hat{x}R}$$
$$\mathbf{y1} := \mathbf{\hat{x}L}$$
$$\mathbf{y0} := \mathbf{y0} + \mathbf{round(y1 \cdot T)}$$
$$\mathbf{y1} := \mathbf{y1} - \mathbf{round(y0 \cdot T)}$$
$$\mathbf{y0} := \mathbf{y0} + \mathbf{round(y1 \cdot T)}$$
$$\mathbf{yL} := \mathbf{-y0},$$
$$\mathbf{yR} := \mathbf{y1}$$

The line with the sign change is to compensate for the sign change of the Lifting implementation of the stereo DCT4 (see above). This is implemented in the function **IntMDCTanafb** for the analysis filter bank, in file **IntMDCTfb.py**,

```
def IntMDCTanafb(x,N,fb):
    #IntMDCT analysis filter bank.
    #usage: y0,y1=IntMDCTanafb(x,N,fb)
    #Arguments: x: integer valued input signal, e.g. audio signal,
    #must be a 2-dim. (stereo) array,
    #2nd index is channel index
    #N: number of subbands
    #fb: coefficients for the IntMDCT filter bank, for the F matrix, np.array with
        1.5*N coefficients.
    #returns y0, y1, consisting of blocks of subbands in in a 2-d array of shape (N,# of
        blocks)

    F0,L0,L1=LiftingFmat(fb)
    D=Dmatrix(N)
    #2 polyphase arrays, 1 for each channel:
    xwin=np.zeros((1,N,int(len(x[:,0])/N)+1,2))
    for chan in range(2): #iterate over the 2 stereo channels
        y=x2polyphase(x[:,chan],N)
        #Lifting step for F, add last (z) dimension for polmatmult:
        y=polmatmult(y,np.expand_dims(F0,axis=-1))
        y=np.round(y) #rounding after Listing step
        y=polmatmult(y,np.expand_dims(L0,axis=-1))
        y=np.round(y)
        y=polmatmult(y,np.expand_dims(L1,axis=-1))
        y=np.round(y)
        y=polmatmult(y,D)
        xwin[:,:,:,chan]=y
    #The IntDCT using multidimensional Lifting:
    y0=xwin[:,:,:,1]
    y1=xwin[:,:,:,0]
    y0=y0+np.round(DCT4(y1))
    y1=y1-np.round(DCT4(y0))
    y0=y0+np.round(DCT4(y1))
```

y0=−y0 #compensate for the sign change of the Lifting implementaion of the stereo DCT4
  (see above)
#test:

For the IntMDCT synthesis filter bank, we take the inverse matrices, in reverse order. Note that in our case of symmetric orthonormal matrices, the inverse matrices are identical with the forward matrices, but our code reflects the more general case, in that it uses the inverse matrices. It is in the function   `IntMDCTsynfb`,

```
#y1=np.round(DCT4(xwin[:,:,:,1]))
return y0[0,:,:], y1[0,:,:]
```

```
def IntMDCTsynfb(y0,y1,fb):
    #IntMDCT synthesis filter bank.
    #usage: xrek=IntMDCTsynfb(y0,y1,fb)
    #Arguments: y0,y1: integer valued subband signals from a stereo file,
    #consisting of blocks of subbands in in a 2−d array of shape (N,# of blocks)
    #fb: coefficients for the IntMDCT filter bank, for the F matrix, np.array with
        1.5*N coefficients.
    #returns xrec, the reconstructed stereo audio signal
    #The synthesis is obtained by inverting each step of the IntMDCTanafb.

    F0,L0,L1=LiftingFmat(fb)
    N=len(y0[:,0])
    Dinv=Dinvmatrix(N)
    #2 polyphase arrays, 1 for each channel:
    xwin=np.zeros((1,N,len(y0[0,:]),2))
    xrek=np.zeros((1,N*(len(y0[0,:])+1),2))
    y0=np.expand_dims(y0,axis=0)
    y1=np.expand_dims(y1,axis=0)
    y0=−y0 #compensate for the sign change of the Lifting implelentaion of the stereo DCT4
    #The inverse IntDCT:
    y0=y0−np.round(DCT4(y1))
    y1=y1+np.round(DCT4(y0))
    y0=y0−np.round(DCT4(y1))
    xwin[:,:,:,1]=y0
    xwin[:,:,:,0]=y1
    for chan in range(2):
        x=xwin[:,:,:,chan] #iterate of the 2 stereo channels
        x=polmatmult(x,Dinv)
        #inverse Lifting steps for the inverse F matrix,
        #add last (z) dimension for polmatmult:
        x=polmatmult(x,np.expand_dims(np.linalg.inv(L1),axis=−1))
        x=np.round(x) #rounding after Listing step
        x=polmatmult(x,np.expand_dims(np.linalg.inv(L0),axis=−1))
        x=np.round(x)
```

Now we can execute the testing routine by calling

`python3 IntMDCTfb.py`

For testing, the number of subbands is $N = 4$, and the program uses a simple ramp signal as input, the numbers from 0 to 63, for the left and also right channel, which has the advantage that it is easily recognized if there is an error in the reconstruction. This signal has a slow rate of change, and hence mostly the lowest IntMDCT subband should be active.

In the beginning, the used Lifting matrices can be seen, also in LaTeX style, then the used test signal of the integer stereo ramp function as a print, then the plots of this input function, the values of the subband samples, and the reconstructed stereo ramp function as print. Observe that the subbands are indeed integer valued.

Still the exact same integers as the input are reconstructed, except for N zeros in the beginning (corresponding to the delay of N samples through the filter bank), and added N samples in the end (Figs. 8.1, 8.2, and 8.3).

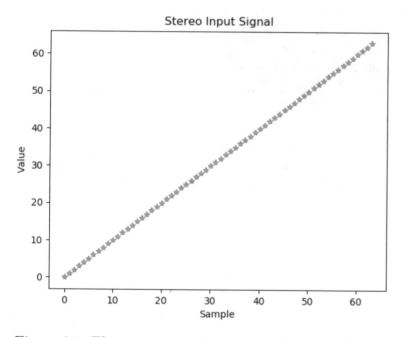

Figure 8.1: The stereo ramp function as input to the stereo IntMDCT, for testing. Each star is representing a sample.

8.3 The Lossless Encoder

We write a program with name "`lossless_rice_audio_encoder.py`", in which we first define the encoded file name of our lossless encoder. It has the same base as the original sound file, but with extension ".lacodrice",

#remove extension from file name:
name,ext=os.path.splitext(audiofile)
#new extension for compressed file:

encfile=name+'.lacodrice'
print("Compressed_file:", encfile)
totalbytes=0

with **open**(encfile, 'wb') as codedfile: *#open compressed file*

We then apply our IntMDCT to the (stereo) audio signal, and apply the Golomb–Rice coder from Sect. 5.3 to its output. For each stereo channel and each subband it computes the Golomb–Rice coefficient b according to Eq. (5.1) and (5.2), and stores them in a length N array `ricecoeff`, which is then stored as side information in our encoded file. For simplicity we take the average of the upper and lower limit for b,

meanabs=np.mean(np.**abs**(ychan[chan,:,:]),axis=−1)
ricecoefff=np.clip(np.floor(np.log2(meanabs)),0,None)

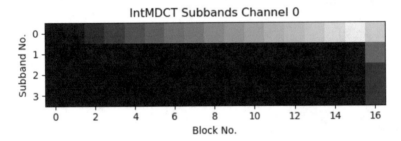

Figure 8.2: The IntMDCT subbands of the ramp function. Dark colours represent small magnitudes, bright colours high magnitudes. Observe that indeed mostly the lowest subband is active, which means the filtering works.

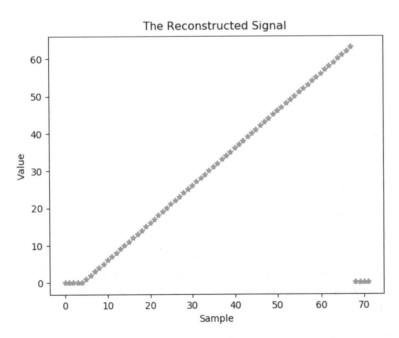

Figure 8.3: The stereo ramp function after reconstruction. Observe that it is exactly the same ramp function, but with additional zeros at the beginning and end.

```
ricecoeffc=np.clip(np.ceil(np.log2((meanabs+1)*2/3)),0,None)
ricecoeff=np.round((ricecoeffc+ricecoefff)/2.0).astype(np.int8) #integer, 8bit
print("ricecoeff=", ricecoeff)
s=struct.pack('b'*int(len(ricecoeff)),*ricecoeff)
pickle.dump(s, codedfile, protocol=-1)
```

Next, using these Rice coefficients b, the subband values are Golomb–Rice encoded and written to the file,

```
for k in range(N): #loop across subbands:
    if (k%100==0): print("Subband:",k)
    #m=2**b
    signedrice=rice(b=ricecoeff[k],signed=True)
    yrice= BitStream(ychan[chan,k,:].astype(np.int32), signedrice)
    #see: http://boisgera.github.io/bitstream/
    #Turn bitstream format into sequence of bytes:
    ys=yrice.read(bytes, np.floor(len(yrice)/8.0))
    totalbytes+= len(ys)
    pickle.dump(ys, codedfile, protocol=-1) #Rice coded subband samples
```

8.4 The Lossless Decoder

The lossless decoder `lossless_rice_audio_decoder.py` simply reverses these steps, first reading the Rice coefficients b from file, then the encoded subband values, and applies the Golomb–Rice decoder,

```
with open(encaudiofile, 'rb') as codedfile: #open compressed file
    fs=pickle.load(codedfile)
    channels=2
    print("fs=", fs, "channels=", channels, )
    numblocks=pickle.load(codedfile)
    numblocks-=4 #last encoded samples might be missing from rounding to bytes
    print("numblocks=", numblocks)
    for chan in range(channels): #loop over channels:
        print("channel_", chan)
        ricecoeffcomp=pickle.load(codedfile)
        ricecoeff =struct.unpack( 'B' * len(ricecoeffcomp), ricecoeffcomp);
        #print("ricecoeff=", ricecoeff)
        ychandec=np.zeros((N,numblocks))

        for k in range(N): #loop across subbands:
            if (k%100==0): print("Subband:",k)
            ys=pickle.load(codedfile) #Rice coded subband samples
            #m=2**b
            signedrice=rice(b=ricecoeff[k],signed=True)
            yricedec = BitStream();
            yricedec.write(ys)
```

ychandec[k,:]=yricedec.read(signedrice, numblocks)
    if chan==0: y0=ychandec
    if chan==1: y1=ychandec

Finally it applies the synthesis IntMDCT and writes the reconstructed signal to file,

print("Inverse␣IntMDCT:")
xrek=IntMDCTsynfb(y0,y1,fb)
xrek=np.clip(xrek,−2∗∗15,2∗∗15−1)
print("Write␣decoded␣signal␣to␣wav␣file␣", decfile)
wav.write(decfile,fs,np.int16(xrek))

To test our lossless audio encoder and decoder, we can record a stereo audio file by entering the following command in our Linux shell,

`arecord -c 2 -r 32000 -f S16_LE -d 5 stereosound.wav`

and speak a few words for the 5 s recording duration. This creates the sound file "stereosound.wav". We can listen to the recorded sound with the command

`aplay stereosound.wav`

Another source of free audio files is [8]. Then we apply the lossless audio encoder with

`python lossless_rice_audio_encoder.py stereosound.wav`

This creates the compressed file "stereosound.lacodrice". We can now compare the sizes of the original sound file with the compressed sound file, using the Linux command

`ls -l stereosound.*`

(the first number for each file gives the size in Bytes). The compressed file should be roughly by a factor of 2 smaller (but keep in mind that this depends on the audio content). Next we can apply our lossless audio decoder,

`python lossless_rice_audio_decoder.py stereosound.lacodrice`

This produces the file with the reconstructed audio "stereosoundlarek.wav". We can take a look at the sizes of the original and reconstructed wave files with

`ls -l stereosound*.wav`

Here we find that the reconstructed file is somewhat smaller. This is because the encoder can only write whole Bytes to file. If the Golomb–Rice coder produces not an entire Byte in the end, the corresponding bits are dropped, in each subband. This is a compromise for the simplicity of our implementation. Hence the last samples are not reconstructed. Since the last blocks are lost, this also means the last reconstructed block of 1024 samples misses its overlap–add from the next block, and hence also those last 1024 samples in the reconstructed wave file are not exactly reconstructed.

We can now compare the original and reconstructed signal, to check if it is really lossless reconstruction. In the Linux shell, we start ipython with

`ipython -pylab`

and then read in the 2 audio files using "scipy.io",

`import scipy.io.wavfile as wav`
`samplerate, audio = wav.read("stereosound.wav")`
`samplerate, audiorek = wav.read("stereosoundlarek.wav")`

Now we can compare the number of samples,

`shape(audio)`
`Out: (160000, 2)`

```
shape(audiorek)
Out: (157696, 2)
shape(audio)[0]-shape(audiorek)[0]
Out: 2304
```

This shows that we lost the last 2304 samples. The IntMDCT leads to a signal delay, just like the MDCT, of 1 Block of N samples, if we have file based processing, where we do not have the blocking delay to assemble samples into Block of length N. Since we chose $N = 1024$ for our lossless coder, the reconstructed signal is accordingly delayed by 1024 samples, for both stereo channels. To check for the exact reconstruction, we need to advance the reconstructed signal accordingly, and remove the last 1024 incorrectly reconstructed samples. Then we can take the difference to the accordingly shortened original audio signal, take the magnitude of the difference, and then their maximum, to check if it is really zero,

```
np.max(np.abs(((audiorek[1024:-1024,:]-audio[:-1024-2304-1024,:]))))
Out: 0
```

We see that it is indeed zero, hence all samples of original and reconstructed within this range, and for both stereo channels, have zero difference, they are indeed identical!

References

1. R. Yu, R. Geiger, S. Rahardja, J. Herre, X. Lin, H. Huang, MPEG-4 scalable to lossless audio coding, in *117th AES Convention* (2004)

2. A. Kalker, I. Shah, Ladder structures for multidimensional linear phase perfect reconstruction filter banks and wavelets, in *Visual Communications and Image Processing* (1992), pp. 12–20

3. W. Sweldens, The lifting scheme: a new philosophy in biorthogonal wavelet constructions, in *Proceedings of the SPIE 2569, Wavelet Applications in Signal and Image Processing III* (1995), pp. 68–79

4. Y. Yokotani, R. Geiger, G. Schuller, S. Oraintara, K. Rao, Lossless audio coding using the IntMDCT and rounding error shaping. IEEE Trans. Audio Speech Lang. Process. **14**, 2201–2211 (2006)

5. R. Geiger, G. Schuller, Integer low delay and MDCT filter banks, in *36th Asilomar Conference on Signals, Systems, and Computers, Pacific Grove* (2002)

6. *Rotation matrix*, https://en.wikipedia.org/wiki/Rotation_matrix. Accessed Sep 2018

7. R. Geiger, Y. Yokotani, G. Schuller, J. Herre, Improved integer transforms using multi-dimensional lifting, in *IEEE International Conference on Acoustics, Speech, and Signal Processing (ICASSP), Montreal* (2004)

8. *Freesound*, https://freesound.org/. Accessed Jan 2019

9 Psycho-Acoustic Pre-filter

9.1 Introduction

The perceptual audio coder we saw uses the same time/frequency resolution for both, the psycho-acoustical model and the data compression. But it does not have to be that both have the same requirement for it. For instance, a high compression ratio demands a higher number of subbands, which then leads to higher system delay of the filter bank. But for communications applications, for instance, teleconferencing with high audio quality, or musicians playing together remotely, a low system delay is required. The latter applications of musicians playing together remotely have the strongest delay requirement. It should be below 30 ms, and for more timing critical instruments at about 6 ms one way delay, for encoding together with decoding. Here it is shown how to use the psycho-acoustic masking threshold to normalize the audio signal to it, and then do any processing in the time domain, or an arbitrary subband decomposition. For instance, to obtain a low algorithmic delay, it would be more beneficial to use predictive coding instead of subband coding.

Or assume that we want to have a pre-filtering stage, such that changes at all frequencies have the same audibility for the ear. This could be done by "normalizing" the audio signal to its masking threshold. After processing we would apply a post-filter, which undoes this normalization.

For both application areas, the low delay coding and the psycho-acoustic pre-filtering, we can use a psycho-acoustic pre- and post-filter. The pre-filter computes the masking threshold of the audio signal at its input, computes a linear filter which has a magnitude transfer function proportional to the inverse of the masking threshold, and applies it to the audio signal. In this domain we would do our processing. After that we apply our post-filter. It receives the parameters representing the masking threshold from the pre-filter, and applies it to its input signal, to undo the normalization of the pre-filter. If we, for instance, apply uniform, frequency flat quantization to the pre-filtered signal, as in the case of the low delay audio coder, then after the post-filter this quantization distortion will be shaped like the masking threshold of the ear, and hence be (hopefully) inaudible to the human ear.

© Springer Nature Switzerland AG 2020
G. Schuller, *Filter Banks and Audio Coding*,
https://doi.org/10.1007/978-3-030-51249-1_9

9.2 Python Example

This principle is described in [1]. Instead of implementing the pre-filter as a predictive filter in the time domain, we can also implement it in the MDCT domain, but at a reduced number of 128 subbands, for a lower delay with a block size of 128 samples. This is the approach we use in our Python example, which also has the advantage that we can re-use some of our functions for our MDCT audio coder.

We start with our program for psycho-acoustic quantization in Chap. 4.8. We use the function "MDCT_psayac_quant_enc", which quantizes the MDCT subband samples according to the masking threshold, but with only 128 subbands, to keep the system delay short. This normalizes the audio signal to its masking threshold. Then we apply the synthesis MDCT to transform these quantized values back into the time domain. This together with the quantization step sizes in array "mTbarkquant" is the output of the pre-filter. The following listing shows the function for the pre-filter:

```
def psyacprefilter(x, fs, quality=100):
    #Psycho-acoustic Pre-filter,
    #Normlizes a signal to its masking threshold
    #Argument: audio signal x, quality (at masking threshold: 100),
    #sampling frequency fs
    #returns: pre-filtered audio signal xpref,
    #masking theshold exponents for the bark subbands mTbarkquant

    N=128 #number of MDCT subbands
    nfilts=64 #number of subbands in the bark domain
    #Sine window:
    fb=np.sin(np.pi/(2*N)*(np.arange(int(1.5*N))+0.5))

    #Analysis MDCT and normalization to the masking threshold and quantization:
    yq, y, mTbarkquant = psyac_quantization.MDCT_psayac_quant_enc(x,fs,fb,N, nfilts,
        quality=quality)
    #Synthesis MDCT, back to the time domain:
    xpref=MDCTfb.MDCTsynfb(yq,fb)
    return xpref, mTbarkquant
```

The post-filter takes the reverse steps. It applies the analysis MDCT to the pre-filtered samples to transform them back into the MDCT domain. Then it de-quantize the subband samples using the quantization step sizes from the array "mTbarkquant" using the function "MDCTsyn_dequant_dec", which then also applies the synthesis MDCT to transform the samples back to the time domain. This de-normalizes the signal from its masking threshold, and is the output of the post-filter. The following listing shows the function for the post-filter:

```
def psyacpostfilter(xpref, fs, mTbarkquant):
    #Psycho-acoustic post-filter,
    #De-normlizes a signal to its masking threshold
    #Argument: pre-filtered audio signal xpref,
    #Sampling frequency fs,
```

#masking theshold exponents for the bark subbands mTbarkquant
#returns: reconstructed audio signal xrek,

```
N=128 #number of MDCT subbands
nfilts=64 #number of subbands in the bark domain
#Sine window:
fb=np.sin(np.pi/(2*N)*(np.arange(int(1.5*N))+0.5))
#Analysis MDCT to the time/frequency domain:
yq=MDCTfb.MDCTanafb(xpref,N,fb);
yq=yq[:,1:-1]; print("yq.shape=", yq.shape) #remove first and last block, which the
    MDCT appended.
#de−normalization to the masking threshold, de−quantization, and MDCT synthesis:
xrek, mT, ydeq = psyac_quantization.MDCTsyn_dequant_dec(yq, mTbarkquant, fs, fb,
    N, nfilts)
return xrek
```

Both, pre- and post-filters are executed one after the other for testing in the main section,

```
if __name__ == '__main__':
    #Example, Demo:
    import sound
    import scipy.io.wavfile as wav

    os.system('espeak_−s_120_"Pre−_and_Post−Filter_demonstration"')
    fs, x= wav.read('fantasy−orchestra.wav')
    #take left channel (left column) of stereo file to make it mono:
    x=x[:,0]
    #fs, x= wav.read('sc03_16m.wav')
    #fs, x= wav.read('test48khz.wav')
    print("Sampling_Frequency=", fs, "Hz")
    plt.specgram(x, NFFT=256, Fs=6.28) #Fs needs to be a float number to avoid error
        message in Python3!
    plt.title('Spectrogram_of_the_Original_Signal')
    plt.show()

    xpref, mTbarkquant = psyacprefilter(x, fs, quality=100)
    plt.plot(mTbarkquant)
    plt.title('The_Masking_Thresholds')
    plt.xlabel('The_Bark_Subbands')
    plt.show()

    xpref=np.round(xpref) # mid tread quantizer
    #xpref=np.floor(xpref)+0.5 #mid rise quantizer

    xrek = psyacpostfilter(xpref, fs, mTbarkquant)

    print("Original_Signal")
    os.system('espeak_−s_120_"Original_Signal"')
```

```
sound.sound(x,fs)
print("Pre−filtered_Signal")

plt.plot(xpref)
plt.xlabel('sample')
plt.ylabel('Value')
plt.title('The_Psycho−Acoustically_Prefiltered_Signal')
plt.show()
os.system('espeak_−s_120_"The_amplified_Pre−filtered_Signal"')
sound.sound(xpref*1000,fs)
print("Reconstructed_Signal_after_Quantization_according_to_the_Masking_threshold")
os.system('espeak_−s_120_"Reconstructed_Signal_after_the_Postfilter"')
sound.sound(xrek,fs)
print("xrek.shape=", xrek.shape)
plt.specgram(xrek, NFFT=256, Fs=6.28) #Fs needs to be a float number to avoid error
    message in Python3!
plt.title('Spectrogram_of_the_Post−Filtered_Signal')
plt.show()
```

Let it run with

`python3 psyacprepostfilter.py`

This first produces the spectrogram of the input signal, as in Fig. 9.1.

Then it plots the pre-filtered signal in the time domain, as in Fig. 9.2.

Observe that the signal after the pre-filter is much smaller from the normalization to the masking threshold. It then plays back the pre-filtered and amplified signal to the sound device. Observe that it sounds like through an equalizer, and distorted. Still after the post-filter the original sound is reconstructed with high fidelity, with perhaps

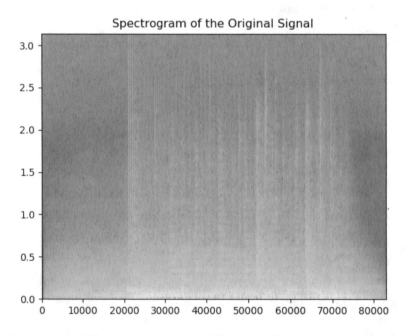

Figure 9.1: The spectrogram of the pre-filtered input signal.

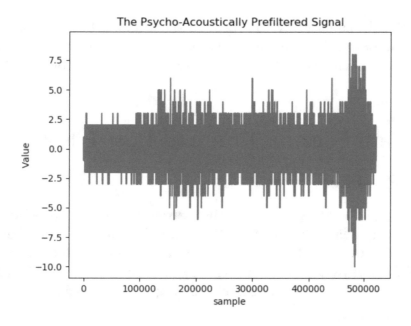

Figure 9.2: The pre-filtered signal.

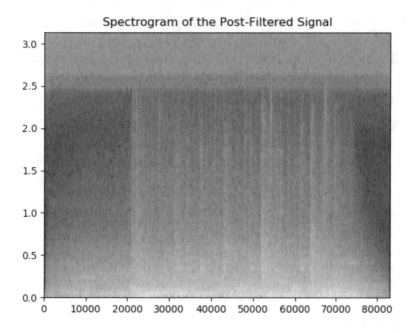

Figure 9.3: The spectrogram of the post-filtered input signal.

some audible artifacts. These artifacts can be reduced by making the quantization step size during pre-filtering smaller, by increasing the "quality" variable. Decreasing the "quality" variable leads to smaller values after the pre-filter, which can be used to obtain a smaller bit rate. The program finally plots the spectrogram of the post-filtered signal, as comparison to the originals spectrogram, as seen in Fig. 9.3.

Observe that it looks basically the same as the original, but with clearly increased noise at the high frequencies. The reason is that the ear has a much reduced sensitivity at high frequencies, such that the system can put quantization noise there and we still do not hear it.

9.3 Separate Files for Pre- and Post-Filter

9.3.1 The Pre-filter

To process the pre-filtered signals it can be convenient to use a pre-filter program to store them separately in audio files, together with their associated masking threshold, such that they can be reconstructed using a separate program for the post-filter. We name this program psyacprefilterToFile.py. For that we use the functions defined in our file psyacprepostfilter.py, which we import in the beginning,

import psyacprepostfilter

When we execute our program in the command line, its first argument is the filename of the audio file to process, and optional the second argument specifies the desired audio quality in percent, where 100% corresponds to the unchanged masking threshold, for above 100% we have a reduced masking threshold, hence potentially improved quality but also potentially increased bit rate; and for below 100% we obtain an increased masking threshold, which leads to a reduced quality but also to potential bit rate savings.

```
if len(sys.argv) < 2:
    print("Usage:_python3_psyacprefilterToFile_audiofile.wav_[quality]")
    print("default_for_quaity_is_100,_higher_number_give_higher_quality_but_higher_bit-rate
        ")

audiofile=sys.argv[1]
print("audiofile=", audiofile)
if len(sys.argv) ==3:
    quality=float(sys.argv[2])
else:
    quality=100.0
fs, x= wav.read(audiofile)
```

Then we construct the two file names, for the pre-filtered audio signal and for the masking thresholds,

```
#remove extension from file name:
name,ext=os.path.splitext(audiofile)
#new extension for compressed file:
preffile=name+'pref.wav'
print("Prefiltered_file:", preffile)
maskingthresholdfile=name+'mT.wav'
print("Masking_Threshold_file_name:", maskingthresholdfile)
```

Now we can compute the masking thresholds and the pre-filtered signal using our functions in `psyacprepostfilter.py`,

```
print("Compute_Prefilter")
for chan in range(channels): #loop over channels:
    print("channel_", chan)
    xchan=x[:,chan]
    xpref, mTbarkquant = psyacprepostfilter.psyacprefilter(xchan, fs, quality=quality)
    print("xpref.shape=", xpref.shape)
    xpref=np.round(xpref) #quantize to nearest integer
    #Convert masking thresholds to 1-D array for storing as audio file:
    mTbarkquantflattened=np.reshape(mTbarkquant, (1,-1),order='F')
    print("mTbarkquantflattened.shape", mTbarkquantflattened.shape)
    mTbarkquantflattened=mTbarkquantflattened[0,:] #remove dimension 0
    if chan==0:
        xprefout=xpref
        mTbarkquantflattenedout=mTbarkquantflattened
    else:
        xprefout=np.vstack((xprefout,xpref))
        mTbarkquantflattenedout=np.vstack((mTbarkquantflattenedout,mTbarkquantflattened
            ))
```

Finally we store both as 8-bit unsigned integer audio files,

```
xprefout=xprefout.T
mTbarkquantflattenedout=mTbarkquantflattenedout.T
wav.write(preffile,fs,np.uint8(xprefout+128)) #write prefiltered audio to 8 bit unsigned integer
    audio file
#(+128 to avoid negative numbers)
wav.write(maskingthresholdfile,fs,np.uint8(mTbarkquantflattenedout))
```

We execute our program using the command line
`python3 psyacprefilterToFile.py fantasy-orchestra.wav 100`
It produces the files `fantasy-orchestrapref.wav` and `fantasy-orchestramT.wav`.

9.3.2 The Post-Filter

We name our post-filter program `psyacpostfilterFromFile.py`. For the post-filter program we have the same first steps for reading in the audio file (this time the pre-filtered audio file), generating the name of the masking thresholds file, and here reading in this masking threshold file. Then we reshape the read in audio file according to the number of channels in the audio file,

```
if len(sys.argv) < 1:
    print("Usage:_python3_psyacpostfilterFromFile_audiopref.wav")
audiofile=sys.argv[1]
print("prefiltered_audiofile=", audiofile)
fs, xpref= wav.read(audiofile)
#remove extension from file name:
```

```
name,ext=os.path.splitext(audiofile)
#remove "pref" from name:
name=name[:-4]
#new extension for compressed file:
postffile=name+'postf.wav'
maskingthresholdfile=name+'mT.wav'
print("Masking_Threshold_file_name:", maskingthresholdfile)
fs, mTbarkquantflattenedout = wav.read(maskingthresholdfile)
try:
   channels=xpref.shape[1] #number of channels, 2 for stereo (2 columns in x)
except IndexError:
   channels=1 # 1 for mono
   xpref=np.expand_dims(xpref,axis=1) #add channels dimension 1
```

Then it loops over the channels to post-filter each channel separately, using our function `psyacpostfilter`, and finally storing the post-filtered signal using 16-bit integer samples,

```
blocks=min(len(xpref[:,0])//N+1, len(mTbarkquantflattenedout[:,0])//nfilts); print("blocks
    =", blocks) #min number of signal blocks in the files
for chan in range(channels): #loop over channels:
   print("channel_", chan)
   print("Compute_Postfilter")
   #subtract the 128 that was added in the prefilter to make it unsigned:
   xchan=xpref[:,chan]-128.0
   #reshape masking thresholds back into a matrix with column length nfilts:
   mTbarkquant=np.reshape(mTbarkquantflattenedout[0:nfilts*(blocks-1),chan], (nfilts,-1),
       order='F')
   xrek = psyacprepostfilter.psyacpostfilter(xchan[0:blocks*N], fs, mTbarkquant)
   print("xrek.shape=", xrek.shape)
   #avoid overflow in wav file by clipping:
   xrek=np.clip(xrek,-2**15,2**15-1)
   if chan==0:
      xpost=xrek
   else:
      xpost=np.vstack((xpost,xrek))
xpost=xpost.T
print("Write_to_Postfiltered_file:", postffile)
wav.write(postffile,fs,np.int16(xpost))
```

We execute our program using the command line
`python3 psyacpostfilterFromFile.py fantasy-orchestrapref.wav`
It produces the post-filtered file `fantasy-orchestrapostf.wav`.
Ideally it should sound the same as the original `fantasy-orchestra.wav`, despite the significant quantization after the pre-filter. If not, the quality percentage can be increased.

9.4 An Ultra Low Delay Audio Coder

We now can take our pre- and post-filters together with our predictive lossless audio coder from Chap. 7 to implement an Ultra Low Delay audio coder [2–4]. Our pre-filter uses an FFT of block length of 256 samples. If we process audio samples as they come in, e.g. from an audio card, this means we have a blocking delay of only 255 samples. But a predictive coder needs no blocking since we use the backward adaptive LMS algorithm for it, hence it adds no delay, only the Entropy coder might need additional delay, as in our example the Rice coder with its coding blocks. To test its compression performance in an **encoder**, we first again apply our pre-filter program to our audio file,
`python3 psyacprefilterToFile.py fantasy-orchestra.wav 100.`
Then we compress the generated pre-filtered file `fantasy-orchestrapref.wav` with our predictive lossless encoder,
`python lossless_predictive_audio_encoder.py fantasy-orchestrapref.wav.`
This generates the compressed file `fantasy-orchestrapref.lacodpred`.
We can now also apply our lossless encoder to the file of the masking thresholds from the pre-filter, for the post-filter,
`python lossless_predictive_audio_encoder.py fantasy-orchestramT.wav`
producing the file `fantasy-orchestramT.lacodpred`.

Now we have 2 files representing the encoded audio file, `fantasy-orchestrapref.lacodpred` (357 kB) and `fantasy-orchestramT.lacodpred` (305 kB). The original, `fantasy-orchestra.wav`, has 2.08 MB, which means we get a compression ratio of about 3. This shows that there is a lot of room for improvement, particularly for the compression of the masking thresholds.

The **decoder** is then taking the inverse steps. First we run the lossless decoder on the encoded masking threshold file,
`python lossless_predictive_audio_decoder.py fantasy-orchestramT.lacodpred.`
producing the file `fantasy-orchestramTlarek.wav`.
Then we need to rename the file for the masking thresholds to the name the post-filter expects,
`cp fantasy-orchestramTlarek.wav fantasy-orchestrapreflmT.wav,`
and we apply the lossless decoder to the encoded pre-filtered file,
`python lossless_predictive_audio_decoder.py fantasy-orchestrapref.lacodpred.`
It produces the file `fantasy-orchestrapreflarek.wav`.
Finally we can apply the post-filter with
`python3 psyacpostfilterFromFile.py fantasy-orchestrapreflarek.wav.`
This produces the decoded file `fantasy-orchestrapreflpostf.wav`.

References

1. G.D.T. Schuller, B. Yu, D. Huang, B. Edler, Perceptual audio coding using adaptive pre- and post-filters and lossless compression. IEEE Trans. Speech Audio Process. **10**, 379–390 (2002)

2. Y. Huang, J. Benesty (eds.), *Audio Signal Processing for Next-Generation Multimedia Communication Systems* (Kluwer Academic Publishers, Dordrecht, 2004)

3. U. Kraemer, G. Schuller, S. Wabnik, J. Klier, J. Hirschfeld, Ultra low delay audio coding with constant bit rate, in *117th AES Convention, San Francisco* (2004)

4. G. Schuller, J. Kovavcevic, F. Masson, V.K. Goyal, Robust low-delay audio coding using multiple descriptions, in *IEEE Transactions on Speech and Audio Processing* (2005), pp. 1014– 1024

10 Conclusions

We saw the fundamentals and implementation examples of audio coding with the help of Python examples. This shows the exact details of its algorithms and systems in a programming language, and should also be the basis of own experiments, because understanding something means being able to create it. We saw the mathematical foundations in the filter bank design and predictive coding, psycho-acoustic basics and their use in psycho-acoustic models. We saw their use in a complete Python perceptual audio coder, a predictive lossless audio coder, and a scalable lossless audio coder, all coder types offered by the MPEG standardization body. This should be helpful for graduate students and engineers working in the field.

© Springer Nature Switzerland AG 2020
G. Schuller, *Filter Banks and Audio Coding*,
https://doi.org/10.1007/978-3-030-51249-1_10

Index

© Springer Nature Switzerland AG 2020
G. Schuller, *Filter Banks and Audio Coding*,
https://doi.org/10.1007/978-3-030-51249-1

Index

Psycho-acoustic model
  bark scale, 127–130
  DFT magnitude, 136
  hearing threshold in quiet, 131
  linear frequency domain, 136
  masking threshold, 136, 139
  non-linear superposition, 134–136
  perceptual evaluation, 140
  and quantization, 140–146
  spreading function matrix, 132–134, 139
  test sounds, 136–140
Psycho-acoustic pre-filter
  application areas, 181
  perceptual audio coder, 181
  post-filter, 187–188
  Python example, 182–186
  separate files, 186–188
  ultra low delay audio coder, 189
Python implementation
  example, 15–20, 47–59, 182–185
  Huffman coder library, 149
  LMS, 118–121
  low delay filter banks, 66–76
  LPC coder, 113–115
  perceptual audio coder, 155–160, 191
  predictive lossless coding, 122–124
  sampling with unit impulse train, 11–14
Python perceptual audio coder
  compressed version, 155
  decoder, 158–160
  encoder, 155–158

Q
Quantization
  channel, 156, 157
  data compression, 181
  de-quantization, 106–107
  error power, 105–106
  mid-tread, 105
  pre-filter, 189
  psycho-acoustic models, 140–144, 182

step size, 106
subband values, 149

S
Sampling
  analog signal, 8–9
  downsampling (*see* Downsampling)
  filter bank, 1
  as multiplication, 5
  perfect reconstruction, 2
  rate, 1, 45, 114
  unit impulse train, 11–14
  upsampling, 21–22
Scalable lossless audio coding, 191
  decoder, 177–179
  encoder, 175–177
  integer-to-integer MDCT, 167–175
Scalefactor-bands, 107, 141
Short time Fourier transform (STFT), 31, 141, 142
Signal to noise ratio (SNR), 82, 83, 94, 106, 120
Simultaneous masking, 94, 132, 134
Spreading function, 132–134, 139
Stochastic gradient descent (SGD), 116, 117
Synthesis filter bank, 27, 34–35
  aliasing, 25
  decoder, 1
  integer values, 167
  MDCT, 57
  output, 40
  reconstructed signal, 95
  subbands signals, 61
  upsampling, 21

T
Time-discrete signal
  Python, 11–20
  zeros removal, 14–15
Time-frequency representation, 2
Time-varying filter bank
  low delay filter bank switching, 96–101

196

Printed in the United States
by Baker & Taylor Publisher Services